Communicating Science, Climate Change and the Environment in Hybrid Media

This volume examines how a new hybrid mediascape represents and contributes to the construction of facts and knowledge in relation to science, environment, and climate controversies, providing a new, critical perspective to the bourgeoning field of science and environment communication.

Arguing that science must be understood from an inclusive perspective, respecting public values and concerns alongside scientific arguments, the authors demonstrate how this will allow us to properly understand the role of science, truth, and factuality alongside the ethical, cultural, and political concerns about science raised in different publics. The chapters focus on the more controversial aspects of science and environmental communication: misinformation, public understandings of science and the environmental crises, vaccination, and the role of the hybrid mediascape in science, environment, and climate conflicts.

Offering a much-needed interdisciplinary approach to understand the role of science of media in science and environment conflicts, this book will appeal to students and academics in the areas of media and communication, journalism, cultural studies, science, environment and risk communication, and digital media studies, as well as sociology and political science.

Mette Marie Roslyng is Associate Professor in Media and Communication at Aalborg University, Denmark. Her research focuses on how discourses of science, technology, and the environment are represented, contested, and politicised in the media and in public debates.

Anna Rantasila is Lecturer in Communication Studies at Lappeenranta University of Technology, Finland. She's also a member of the Disinformation, Propaganda & Soft Power Research Lab at LUT. Her current work focuses on affect and emotion in various online environments, particularly in the context of news, disinformation, and popular culture.

Anna Maria Jönsson is Professor in Media and Communication Studies at Södertörn University, Sweden. She has extensive experience of research about journalism and climate change, science communication, and public engagement.

Routledge Studies in Media, Communication, and Politics

28 Refugee Settlement in Australia
A Holistic Overview of Current Research and Practice
Aparna Hebbani

29 Media, Dissidence and the War in Ukraine
Edited by Tabe Bergman and Jesse Owen Hearns-Branaman

30 European Media Systems for Deliberative Communication
Risks and Opportunities *Edited by Zrinjka Peruško with Epp Lauk and Halliki Harro-Loit*

31 Media Tactics in the Long Twentieth Century
Edited by Marie Cronqvist, Fredrik Mohammadi Norén and Emil Stjernholm

32 State-Sponsored Disinformation Around the Globe
How Politicians Deceive Their Citizens *Edited by Martin Echeverria, Sara García Santamaría and Daniel C. Hallin*

33 Kremlin Media Wars
Censorship and control since the invasion of Ukraine
Edited by Wendy Sloane and Aleksandra Raspopina

34 Contestation in Civil Society
Noise, Listening and Civil Action
Stefanie Pukallus

35 Communicating Science, Climate Change and the Environment in Hybrid Media
Constructed Facts, Contested Truths
Edited by Mette Marie Roslyng, Anna Rantasila and Anna Maria Jönsson

www.routledge.com/Routledge-Studies-in-Media-Communication-and-Politics/book-series/RSMCP

Communicating Science, Climate Change and the Environment in Hybrid Media

Constructed Facts, Contested Truths

Edited by Mette Marie Roslyng, Anna Rantasila and Anna Maria Jönsson

Routledge
Taylor & Francis Group
NEW YORK AND LONDON

First published 2025
by Routledge
605 Third Avenue, New York, NY 10158

and by Routledge
4 Park Square, Milton Park, Abingdon, Oxon OX14 4RN

Routledge is an imprint of the Taylor & Francis Group, an informa business

ISBN: 978-1-032-76665-2 (hbk)
ISBN: 978-1-032-76668-3 (pbk)
ISBN: 978-1-003-47955-0 (ebk)

DOI: 10.4324/9781003479550

Typeset in Times New Roman
by Newgen Publishing UK

Contents

Figures and tables

Figures

Tables

Author biographies

Ernesto Abalo is Associate Professor in Media and Communication at the School of Education and Communication, Jönköping University, Sweden. His research interest revolves around media, discourse, and ideology, and his research centres on critical discourse analysis. He has studied a range of topics, such as the mediation of environmental problems, news media construction of drugs, international journalism about Latin America, and constructions of class in different media. Currently, he studies the user interfaces of Swedish digital welfare services in a research project on older adults and welfare technologies.

Óscar García Agustín is professor of Democracy and Social Change at the Department of Culture and Learning at Aalborg University, Denmark. He is head of the DEMOS research group. His main research areas are populism, democracy, social movements, migration, and solidarity. He was PI of the project 'Geographies of Populism in Europe' (2020–2023) funded by the Danish Research Council. With Marco Briziarelli, he has co-edited *Podemos and the New Political Cycle* (2018). He is author of *Sociology of Discourse: From Institutions to Social Change* (2015) and *Left-Wing Populism. The Politics of the People* (2020).

Gregers Andersen is External Lecturer at the Department of Communication and Psychology, Aalborg University, Denmark. His research is broadly situated within the environmental humanities, as he examines how sustainable and unsustainable futures are envisioned in a range of communicative forms, stretching from the future speculative worlds of literary and cinematic fictions (cli-fi) to the sustainable and unsustainable futures imagined and described by political parties and companies. He is the author of *Climate Fiction and Cultural Analysis. A New Perspective on Life in the Anthropocene* (Routledge, 2020).

Massimiliano Demata is Professor of English Linguistics at the Department of Culture, Politics and Society of the University of Turin, Italy. His research interests include discourses of nationalism and borders, conspiracy theories, the language of populism, and social media discourse. He has held research and teaching positions in France, Germany, Israel, Luxembourg, Sweden, the UK, and the USA. His latest work includes *Discourses and Borders and the Nation in the USA* (Routledge 2022) and the multi-authored book *Voices of Supporters* (John Benjamins 2023). He is also one of the co-editors of the *Routledge Handbook of Discourse and Disinformation* (Routledge 2024).

Olga Dovbysh is a postdoctoral researcher at the Aleksanteri Institute, University of Helsinki, Finland. Her research focuses on media and internet control in restricted media regimes, journalistic cultures, and environmental and climate communication. She is a co-founder of the academic initiative 'Online Talks on Eurasian Media,' aimed at bringing together researchers and journalists to discuss current and emerging problems in the field of media and communication across the vast territory of Eurasia. Dovbysh's recent publications have appeared in journals of *Journalism*, *Digital Journalism*, *Media & Communication*, and *Social Media and Society*.

Isabel Jerne has researched how geography and space shape political ideas, particularly through the lenses of populism and counterpublics. While working as a research assistant on *The Geographies of Populism* project, she explored the impact of spatial and environmental dynamics within political landscapes.

Anna Maria Jönsson is a Professor in Media and Communication Studies at Södertörn University, Sweden. She has extensive experience of research about journalism and climate change, science communication, and public engagement. Her recent publications address topics such as news media's portrayal of research, public trust in research, media reporting and trust concerning the Covid pandemic, and how data journalism can contribute to civic engagement.

Gorm Larsen is an Associate Professor at the Department of Communication and Psychology, Aalborg University, Denmark. He has a strong foundation in semiotics and narratology, focusing on the act of narration within Bakhtin's conceptual framework. In recent years, this research has also included communication from groups that position themselves in opposition to power and the elite. Another main area lies in the concepts of shame and guilt. He has explored these themes in literature and film, as well as in sociological contexts, investigating how guilt and shame have changed historically and how they can be viewed as a seismograph for our culture.

Maarit Mäkinen works as a university researcher at Tampere University, Finland. She has experience in multidisciplinary research and development projects, focusing on themes such as participation, communication, media literacy, and digitalisation. Her latest research project, 'Fair Media: From Silencing to Respectful Plurality,' examined the opportunities and barriers to citizen media participation and inclusive media culture.

Mika Perkiömäki is a postdoctoral researcher at the Aleksanteri Institute, University of Helsinki, Finland. His areas of interest include environmental humanities, Russian media studies, Russian environmental journalism, ecocritical studies of Russian literature and film, nuclear fiction, and mediated climate discourses. Perkiömäki is the Editor-in-Chief of *Idäntutkimus*, the Finnish review of East-European studies, and the vice-chair of the Society of Finnish Slavists. His recent publications have appeared in journals such as *Journalism*, *The Soviet and Post-Soviet Review*, and *Slavonica*.

Anna Rantasila is a Lecturer of Communication Studies at the School of Engineering Sciences at Lappeenranta University of Technology, Finland. She's also a member of the Disinformation, Propaganda & Soft Power Research Lab at LUT. Her current work focuses on affect and emotion in various online environments, particularly in the context of news, disinformation, and popular culture. She's also interested in questions of trust and scientific expertise. In her 2020 doctoral dissertation, she studied online news comments about the 2011 Fukushima Daiichi nuclear disaster.

Mette Marie Roslyng is Associate Professor in Media and Communication at Aalborg University, Denmark. Her research focuses on how discourses of science, technology, and the environment are represented, contested, and politicised in the media and in public debates. She is particularly concerned with green citizenship, citizen engagement, and the role of digital and alternative media in science debates. Her latest research interests include work on democratising green conflicts and affective politics.

Gabor Sarlos is an Associate Professor at the University of Roehampton, UK. Before joining, he was Discipline Lead of the BA in Communication program at RMIT University Vietnam. Previously, he held various teaching positions in the UK and Hungary. His academic interests and activities are related to public perception of energy and climate change, risk and crisis management, environmental awareness, and the development of sustainability communication models. Before moving to academia, he accumulated two decades of experience in developing communication strategies and campaigns for companies and in the third sector.

Virág Vécsey is Assistant Professor at the Institute for Art Theory and Media Studies at the Eötvös Loránd University, Budapest. Her fields of research are animation studies, environmental communication, and ecocriticism. Her doctoral research explores the temporal changes in human-nature relationships as represented through European animation, in context of the changing social, political, and industrial landscape. She is founder and head of the BA media design specialisation at the Department of Media and Communication at ELTE.

Preface

As researchers, we have long been concerned with the status of scientific and environmental discourse in the media. Through our involvement as chairs of the Science and Environment Communication Section of the European Communication Research and Education Association (ECREA) and the Nordmedia Division in Environment, Science and Risk Communication, we have followed the scholarly developments on the topic for quite a while. We already explored the question of how facts and truths are constructed in the media in a collaborative workshop at the MeCCSA (Media, Communication and Cultural Studies Association) conference. This event took place in January 2020 in Brighton, UK, and a few weeks later, the Covid-19 pandemic brought along a period of general chaos that disrupted what we knew as everyday life. The pandemic brought to the forefront the issues of trust in science, science contestations, and visions of how modern life is framed by technologies, relations between people and animals as well as markets and authorities. Furthermore, climate change had entered the global political agenda, driven by citizen activism and young people's movements in a way that surpassed the civic engagement of previous decades.

The topic of this edited volume, *Communicating Science, Climate Change and the Environment in Hybrid Media: Constructed Facts, Contested Truths* is therefore even more timely than ever in the wake of the Covid-19 pandemic, climate change still being a highly urgent issue, and public discourse in general is becoming more polarised.

Several of the authors in this volume have been part of the project since 2020, and we are very grateful to all the authors for their important contributions and their patience with the process.

We would also, in particular, like to thank Kasper Kotisaari for his work on the index and LUT Department of Social Sciences for supplying the funds for this invaluable task. Finally, we want to thank Suzanne Richardson and Saraswathy Narayan at Routledge for encouragement and for guiding us through the process of finalising this book.

November 2024, Aalborg, Stockholm, and Lappeenranta,
Mette Marie Roslyng, Anna Rantasila, and Anna Maria Jönsson

1 Introduction

Contesting truths in science and environment communication

Anna Maria Jönsson, Mette Marie Roslyng, and Anna Rantasila

A recurring and dominant theme in public debates is how to understand and talk about controversies relating to science and the environment. As the Covid-19 pandemic, climate change, environmental degradation, and controversial new technologies become pertinent on the global political agenda, new and old dilemmas of how humans interact with nature, technologies, capital, and each other are brought forward. This puts into question well-known as well as new quandaries about the current and future role of science and scientific expertise in society. On the one hand, political actors rely on science to produce the facts and evidence required as inputs in decision-making. On the other hand, the privileged position of science to provide fulfilling answers to scientific and environmental predicaments is increasingly challenged in the public domain when faced with scientific uncertainty, complexity, disagreement, and reduced levels of trust. This anthology addresses the myriads of dilemmas arising from the contestations of scientific and environmental facts in public spheres in recent debates and science controversies. We are, in particular, intrigued by the question of how the media contribute to the construction of facts and knowledge in these conflicts. The media is thus central to understanding the role and status of scientific truths as they unfold in public spheres.

In the hybrid media environment, no representation is created in a vacuum (Chadwick, 2013). For example, journalistic narratives influence fictional accounts of science and technology and social media content, which in turn affect the way these events and issues are framed in journalism. Moreover, as Valdivia (2013) notes, media studies can be described as an "interdiscipline" as its field of study encompasses a multitude of media, and the theories and methods of the discipline often overlap with other disciplines, and the boundaries between them appear often porous (Rantasila, 2020). Therefore, we find it only fitting that the interdisciplinary nature of this volume entails that the definition of "media" used here is also relatively broad and varies slightly in each chapter. The empirical studies of the chapters cover a wide variety of sources, including social media platforms such as Instagram, Facebook, or X (formerly Twitter); journalistic publications and their digital editions such as local and global newspapers and national public broadcasters; as well as entertainment media such as fiction film and television series. Thus, depending on the chapter, "media" in this volume can refer to a

DOI: 10.4324/9781003479550-1

certain *medium*, such as film, television, or newspapers; it can refer to *a set of practices and practitioners*, such as in "journalistic news media"; or it can be used to refer to the *content* shared on various platforms and publications; or, finally "the media" can also refer to *the social and cultural institutions*, such as film industry or journalism (Rantasila, 2020). What connects these different definitions and understandings of the term and the forms of mediated expression and information delivery is that they all provide different representations of events, phenomena, and actors related to things that also concern science, environment, and technology. By examining these different modes of representation together in this volume, we can provide deeper insights into contemporary debates around science and environmental issues.

The changing role of the hybrid media environment (Chadwick, 2013) in conveying and constructing current debates of the role of science in society has become an object for science and environmental communication scholarship (Pinto et al., 2019; Lewis et al., 2021). Hybrid media systems refers to the ongoing mutual adaptation of structures, norms, logics, and practices of the legacy media like mainstream journalism, with those of internet-native media, social media, and mega-digital platforms. This hybrid situation also can affect power structures, and Jungherr, Posegga & An (2019) launch the concept of *discursive power* as the ability of actors to harness the logics of legacy and social media to dominate the discourse.

The notion of discursive power becomes visible through developments in social and digital media that have raised the issues of factuality and truths into the public debate in relation to phenomena such as post-truth, fake news, and misinformation (Grech, 2021; Fuchs, 2020). Mis- and disinformation are, most commonly, contrasted with some form of evidence-based knowledge (Vraga & Bode, 2020, p. 136) and thus understood as information that is either false, misleading, or manipulative. Based on this distinction, some authors emphasise the role of professional journalism in upholding a certain standard of public science communication through activities such as fact-checking (Luengo & García-Marin, 2020). Meanwhile, other studies focus on how media sets conditions for the development of science populism or counter-knowledge positions (Eslen-Ziya & Giorgi, 2022; Mede & Schäfer, 2020) as well as conspiracy theory (Demata, Zorzo, & Zottola, 2022). Post-truth has emerged as a central concept in this debate to denote a situation in which public opinions are based less on facts and more on emotive or personal beliefs. According to Speed and Mannion, the idea of post-truth has roots in a situation of hyperreality described as "the inability to distinguish the real from the false" (Speed & Mannion, 2017, p. 250). Post-truth has thus consistently been used to understand political populism (e.g., Suiter, 2016). While authors in the field either endorse or take issue with the notion of post-truth (Portelli & Oladi, 2021; Büscher, 2021; De Cleen, 2018), the question of how to make sense of the circulations of conflicting ideas about facts in current public debates remains largely unresolved.

Misinformation and conspiracy theories undoubtedly pose a democratic challenge to public and political coherence. However, defining misinformation in contrast to evidence-based knowledge poses several challenges due to the bounded nature of both popular and expert-based knowledge (Vraga & Bode, 2020). Both forms are marked by different kinds of scientific uncertainty, knowledge gaps, and disagreement. Therefore, in this book, we want, first, to avoid the pitfall of either distinguishing between illiberal/liberal or repressive/progressive political groups and actors, or entering a normative discussion of political projects promoted by the media. Second, rather than taking a starting point in the relationship between authoritative science produced and communicated by, e.g., government authorities or mainstream media on one side, and alternative news production on the other, we explore how media and political logics intertwine and unfold, thus allowing us to better understand how these particular kinds of public spaces develop and how they produce scientific discourses. Following Tuukka Ylä-Anttila, we suggest that movements critical of mainstream science produces discourses of *counter-knowledge* defined as: "alternative knowledge which challenges establishment knowledge, replacing knowledge authorities with new ones, thus providing an opportunity for political mobilization" (Ylä-Anttila, 2018, p. 359). Counter-knowledge thus can produce misinformation but does not automatically do so.

These issues have taught us the importance of public understandings of science, particularly as facts in science and environment conflicts are seen as co-constructed by different public actors. As forcefully formulated by Sheila Jasanoff: "Scientific knowledge ... is not a transcendent mirror of reality. It both embeds and is embedded in social practices, identities, norms, conventions, discourses, instruments and institutions" (Jasanoff, 2004, p. 3). Drawing only on expert and government elite positions as authoritative voices can lead to public dissent, science populism, and polarisations. We argue that science must be understood from an inclusive perspective, respecting public values and concerns alongside scientific arguments. Furthermore, a multifaceted discussion and even disagreement is necessary for a healthy democracy (Fraser, 1990; Mouffe, 1999), and this also applies to science-related topics that often tend to be delineated as the territory of scientific expertise. Adopting an inclusive approach in science, environment, and climate communication will not devalue science. On the contrary, this position allows us to properly understand the role of science, truth, and factuality alongside the ethical, cultural, and political concerns about science raised in different publics.

Exploring the role of the rapidly changing field of digital media, and particularly that of social media platforms from this perspective, thus invites inquiry not only of mis- and disinformation but also discussion on public understanding of science and environmental crises. Moreover, this guides our interest into the wider question about the role of the hybrid media environment in science, environment, and climate conflicts. Furthermore, the Covid-19 pandemic lifted several phenomena and issues central to the titular theme of this volume, such as conspiratorial thinking, contesting science, and suspicion towards new technologies, to the popular consciousness and public discussion more vividly than perhaps ever before. However, we would like to note that none of these phenomena are neither particularly new nor are they symptomatic of the 2020s. As some of our chapters note, conspiratorial thinking remains part of climate change denialist logic (e.g., Vécsey, 2025), and scepticism towards scientific experts and new technologies has been a part of discussions about nuclear energy since the 1960s (Rantasila, 2025).

We argue that public understandings of science and environmental issues involve discussions and contestations in the media over what can be considered as facts and truths. This concerns not only different interpretations of science but also how science is given meaning culturally and politically and the various roles different types of media play in this process. The purpose of this book is thus to address the popular misconceptions that mainstream news media either provides a biased or distorted view of science and environmental issues, or that the media is an infallible broker of indisputable facts (Goldenberg, 2020). This leads us to a better understanding of the role of media in the cultural, social, and political construction of facts on global risks, new technologies, and environmental challenges. Representations of science in conflicts regarding climate change and the environment take place in public spheres that can be seen as antagonistic and political (Mouffe, 2005) as they concern fundamental issues regarding resource distribution, environmental responsibility as well as environmental justice and global and social solidarity. In fact, the construction of a frontier between (illegitimate) populist science and (democratic and legitimate) public engagement in science is indeed one of these antagonistic lines of demarcation that is both reproduced and contested in different types of media (Roslyng, 2022). Several chapters in this volume address how science and the environment become part of a political struggle in counter-hegemonic or counter-public spheres (Larsen & Roslyng, 2025; Agustín & Jerne, 2025).

Whose truth? Questions of definition in the age of contested science

At the heart of this volume are questions of truth and facts, and the contest over who gets to define these terms. As controversies about, for example, the connection between fossil fuels and climate change or the Covid-19 pandemic illustrate, while scientific facts may be indisputable on the level of empirical evidence and scientific discourse, various actors in the hybrid media system compete on the ability to be the ones who *define the interpretation* of those facts and what is considered as relevant facts (e.g., Alinejad & Van Dijk 2023; Van Dijk & Alinejad, 2020). As Van Dijk and Alinejad (2020) note, contemporary social media platforms and their discursive polyphonies challenge traditional models of science communication. In other words, the definition of a scientific institution on an issue may have been relatively uncontested in the public sphere in the 1980s because of the relatively well kept gates of mainstream news media. However, with the advent of the Internet and social media, competing definitions have gained more visibility and traction as the number of voices and actors in the hybrid media environment has increased (Van Dijk & Alinejad, 2020).

Normativity and questions of legitimacy often enter into the discussion about truth, particularly when discussing science and scientific phenomena. After all, science can be understood as the pursuit of truth about the world in a most profound sense (e.g., Kuula, 2006). However, as communication and media studies scholars, we recognise that the question of truth is far from simple, as there are ways of producing legitimate knowledge that sometimes fall outside the quantifiable or the discursive (see, e.g., Crasnow & Intemann, 2020; Rolin, 2020). Moreover, facts have a social history, but acknowledging this does not downgrade their epistemic status (Goldenberg, 2020). Therefore, it is possible to simultaneously understand that climate change is an actual, ongoing phenomenon, and to recognise that scientific knowledge about climate change is produced in a specific social and cultural setting. In this volume, we seek to make these public contests over definitions of scientific truth visible, as they often are played out in the popular media as much as in the realms of scientific debate. These contests over definitions of scientific truth are also tied into a broader development of increasing scepticism towards traditional social institutions, such as mainstream news media and democratic political organisations (e.g., Alinejad & Van Dijk, 2023; Carlson, 2018; Fawzi, 2018; Holt, 2018; Jönsson, 2020). This is often linked to an increase in political populist movements and rhetoric (e.g., Tuomola, 2023, Hatakka & Herkman, 2022), as several chapters in this volume also illustrate. By highlighting these contexts, we can also provide ways to understand why some tend to support a version of truth that may not be as factual as the other versions.

Controversies about scientific claims are often framed as a battle between the informed versus the uninformed, science versus anti-science (Goldenberg, 2020), and this stance is sometimes implicitly also visible in the chapters of this volume. However, as the multiple examples discussed in this volume point out, the empirical reality is rarely as clear and clean cut. As Goldenberg (2020) and Goldenberg et al. (2023) argue, phenomena such as vaccine hesitancy are often linked to mistrust instead of simple lack of information. She suggests that instead of a problem of an uninformed public who does not know "the truth", scepticism and mistrust towards scientific knowledge and institutions is a problem of scientific governance (Goldenberg, 2020). In order to nurture trust in the scientific truth, the scientific establishment has to be trustworthy. Ridiculing and labelling those who are sceptical of scientific knowledge and institutions as unhinged is rarely productive, as there often are legitimate concerns behind even the extreme forms of this distrust.

Adopting a discursive approach to scientific and environmental factuality, as seen in several of the chapters in this book, redirects attention away from truth *per se* towards a focus on how truth-claims are performed in the media. In the context of post-truth politics, Sorensen and Krämer (2024) convincingly argue that public truth-claims are increasingly seen as performances of authenticity rather than objectivity. This is not only the case in polarised or populist political situations but also in deliberative politics where truth-claims are debated and scrutinised in public spheres (Sorensen & Krämer, 2024). The discursive construction of scientific authority claimed by, e.g., experts and governments take the form of articulations of truth-claims in a way that is decidedly hegemonic (Roslyng & Larsen, 2021). Discourses of counter-knowledge, on the contrary, engage in the construction of truth-claims that contest and rearticulate discourses of authoritative science and present alternatives to these as explored by several chapters in this volume (Mäkinen, 2025; Demata, 2025; Larsen & Roslyng, 2025; Agustín & Jerne, 2025).

Environmental and climate truths in media

This volume is divided into three thematic sections. The four chapters in section one, *Environmental and climate truths in media*, discuss how public and mediated discourse on climate change and environmental issues are arenas of fierce contest over facts, truthfulness, and the righteous interpretation and relevance of the aforementioned facts. Each of the chapters take on a different empirical perspective. The first chapter discusses coverage of an environmental issue in mainstream news media, while the second addresses how visual social media is used to disseminate denial of climate change. The third chapter of this section explores how anthropogenic climate change is addressed in fiction films and literature in the West. The fourth and final chapter returns to journalism and politics and discusses how populism and environmental issues are tied together.

Together, the chapters provide an overview of how science and science-related phenomena permeate contemporary society and media. As Weignart (2022) notes, the past 50 years have seen an expansion in public communication about science. In addition, Alinejad and Van Dijk (2023) note that news coverage of climate change alone has increased noticeably in the past 20 years. Thus, it is not surprising that these topics seep into vernacular media and fictional works as well, as Chapters 2 and 4 illustrate. However, increased mediated presence of science-related topics, such as climate change, also increases the risk of what Alinejad and Van Dijk (2023, 30) call "epistemic bubble-formation". One example of this phenomenon are the social media posts of climate change denialists discussed in Chapter 3.

In the first chapter of the section, Chapter 2, "The scientification of risks and the risks of scientification: Insights from the coverage of artificial turf pitches as microplastic pollutants in Sweden", Ernesto Abalo explores the notion of scientification and how it can be used to understand how science in specific discursive strategies in news media shapes the reporting around environmental risks. Abalo discusses how science, through the use of specific discursive means, becomes a key component in the construction of different arguments around environmental risks. He explores this dynamic through a critical discourse analysis of the media coverage of artificial turf pitches as microplastic pollutants in Swedish aquatic environments. According to Abalo's findings, the use of scientification unveils a dynamic and complex relationship between science, environmental risks, and media discourse.

In Chapter 3, "Web of denial. Climate change denial discourse on Instagram" by Virág Vécsey, critical discourse analysis and content analysis are brought to bear in an analysis of the visual elements of climate change denial in Instagram, on posts found under #climatehoax or #climatechangehoax. According to Vécsey, climate denial communication on Instagram is characterised by scepticism about anthropogenic climate change embedded in a broad web of conspiracy theories, forming an incomplete and self-contradictory super conspiracy theory. Content posted under the examined hashtags re-frames climate change-related visuals used and established by the mainstream media to follow the denier narrative. Rhetorically, humour and irony are the main weapons of the climate change denialist discourse on Instagram, reflecting existing findings about the use of memes and other forms of humour by the far-right.

In Chapter 4, "Cli-fi and five narratives of future warming", Gregers Andersen provides another angle to mediated discourse on climate change by analysing Western fiction novels and feature films about anthropogenic climate change. From a substantial body of literature and film, Andersen identifies five narratives that dominate the genre of climate fiction and explore possible futures where the Earth's environment and climate are irreparably altered. In addition to discussing the features of these narratives in detail, Andersen points to the relations between these templates and some of the ancient myths that have played a key role in the cultural history of the West. According to Andersen, the genre of climate fiction provides a valuable way to examine possible, if rather bleak futures, ways of coping and possibilities of repairing the damage done to the planet.

Moving back to the realm of journalistic and political discourse on environment and climate change, Óscar García Agustín and Isabel Jerne discuss how populism and environmental issues collide in Chapter 5, "Green populism: Counterpublics and the formation of counterknowledge". Agustín and Jerne conceptualise the public sphere on climate transformation as an arena of contestation, where movements, individuals, and groups challenge each other and articulate their perspectives and interests either by recognising the legitimacy of others or by eliminating others. Based on their analysis of both right-wing and left-wing performances of climate activism, Agustín and Jerne claim that "green populists" shape counterpublics that, besides reinforcing their identity (as green populist movements), enhance the formation of "green knowledge" and activate "repertoires of knowledge practices" to produce change.

Contested science: Conspiracy and counter-knowledge

The second section of this volume, "Contested science: Conspiracy and counter-knowledge", delves deeper into themes that many of the chapters in the first section also touch upon, namely those of conspiratorial thinking and contesting established scientific knowledge. After an initial hype on the democratic potential of social media, certain platforms, particularly Twitter/X, Facebook, and Tik Tok, have been scrutinised for their role in enabling the spread of conspiratorial claims during the recent decade or so (e.g., Ferrara, Cresci, & Luceri 2020; Schatto-Eckrodt et al., 2024). As Schatto-Eckrodt, Clever, and Frischlich (2024) note, exposure to conspiracy theories tends to decrease trust in democratic institutions and legitimate forms of political participation while simultaneously fostering tolerance for illegitimate forms of political participation. In addition to posing a threat to democratic institutions and practices, conspiracy theories and conspiratorial thinking also challenge established education, science, and medicine. Chapters 6 and 7, which both focus on conspiratorial claims around the Covid-19 vaccinations, illustrate how claims that mostly circulate in social media also bleed into mainstream media discourse. Both chapters also shed light on how participants in these debates construct the other side as a polar opposite, thus accelerating the polarising developments further.

The rise of polarisation of opinion has not been limited to online spaces, as the many on-site demonstrations and political debates around climate change and Covid-19 vaccines and policies, for example, across the globe during the 2020s illustrate. Each of the chapters in this section addresses both the role of social media and the general rise of polarisation in the context of the Covid-19 pandemic but through different empirical material. Chapters 6 and 7 focus more on the interplay of social media and mainstream media, while Chapter 8 provides an interesting take on how mainstream news media frames those who take their opinions to the streets.

In Chapter 6, "Fighting (for) truth? Alex Jones, the WHO and the legitimation of conspiracy discourse", Massimiliano Demata analyses the controversial public figure's online videos about the origins of the Covid-19 respiratory virus. Demata argues that far from being purely emotional and removed from facts or science, conspiracy discourses employ certain persuasive strategies that evolve around the authority of scientific sources and evidence. In addition, his analysis points out how conspiratorial media figures like Jones use visual and rhetorical tactics of validation that highly resemble those of established science and mainstream media, producing what Demata calls "epistemic masquerade".

Chapter 7, "Knowledge and counter-knowledge: The construction of facts in the vaccination debate" by Gorm Larsen and Mette Marie Roslyng, focuses on the analysis of Danish news media and Facebook group debates about Covid-19 vaccinations during the lock-down period in 2020. In their chapter, Larsen and Roslyng conduct a qualitative discourse and linguistic analysis of how hegemonic and counter-hegemonic positions articulate competing truths about vaccine risks while drawing on facts – and ideas – from different and often opposing perspectives. Based on their analysis, they examine if media representations of science conflicts can be seen as polarising and counter-hegemonic or if they support a hegemonic authoritative knowledge voice.

In the final chapter of this section, Chapter 8, "Citizen activists or pandemic deniers? Alternative voices in the Finnish journalistic media during the COVID-19 pandemic", Maarit Mäkinen examines if the Finnish mainstream media showed a tendency to exclude critical views of citizens and possibly engaged in communication styles that strengthened polarisation during an exceptional time. By analysing news articles and a documentary series about so-called pandemic deniers, Mäkinen argues that in their coverage of alternative voices, the Finnish mainstream media tended to include elements that increased polarisation of opinion, such as lack of dialogue, justification with speaker categories, categorising people as well as using extreme expressions. In addition, the material showed fact construction based on a selected moral code, which can increase polarisation between people.

Constructing public knowledge and trust

The third and final section of the book moves on to discuss how media deals with complex scientific topics, how it discursively constructs knowledge and counter-knowledge, and if and how these dealings impact public knowledge and trust in science. Scientific knowledge is today an important part in democratic citizenship, and knowledge as well as trust are important factors for forming public attitudes, behaviour, and engagement (Davis & Horst, 2016; Goldenberg et al., 2023).

This volume does not empirically analyse public knowledge about science, but previous research has shown the importance the media has for the public in defining what is right and relevant knowledge (Davis & Horst, 2016). In the centre of this research is also the role of the media providing information to the public that is fundamental for knowledge creation. Although the media is not the only actor and factor affecting public knowledge about science, the role of media for constructing knowledge and how this is perceived and used by the public is fundamental for democracy (Davis & Horst, 2016).

The relation between science communication and public knowledge about scientific issues has a long research tradition, albeit the perspectives and focus have shifted over the years regarding for example why public knowledge of science is considered important and what can be done to improve it. Up until a couple of decades ago, research focused mainly on how to solve the perceived problem that citizens' knowledge levels were too low and that the opinions about science were too negative (science literacy model, public understanding of science). Since the 1990s though, focus has shifted to the importance of promoting public trust and dialogue with citizens about research issues and results (Cunningham-Burley, 2006; Schäfer, 2009).

Trust is a broad concept with several dimensions, both related to interpersonal relationships and to societal institutions and actors. According to Goldenberg et al. (2023), trust affects almost all kinds of human interaction and can be described as the "glue of society". Trust is always about relations, and for this volume, it is the relationships between subjects (individuals) and societal objects and institutions that is most important. Epistemic trust is a concept focusing on trust in scientists and their competence and knowledge and puts the scientific expert in focus. According to previous research, this kind of trust increases through education, transparency, and accountability (Hendriks et al., 2016; Lynch, 2016; Sulik et al., 2021).

It has often been stated that there is a polarisation in public discourse in relation to what is considered relevant and correct facts and information. This polarisation seems to be related both to forms of political systems, where authoritative systems generally question societal knowledge institutions, and to individual factors. Previous research shows that levels of trust in science and scientists in general, as well as in different disciplines, differ between population groups (Achterberg et al., 2015; Bergström & Oscarsson, 2015). There also seems to be a relation between level of knowledge and level of trust even though this relation is not unambiguous. Research about public trust in science in Sweden indicates an increased polarisation between the groups with high respectively low levels of trust (Jönsson, 2020).

We see today a trust polarisation not only in relation to science but to public knowledge institutions, like media and higher education, in general. This has been very clear in relation to wicked problems and societal crises as for example climate change and the Covid pandemic (Agly, 2020; Bromme et al., 2022; Sulik et al., 2021), and the issue calls for urgency since public trust in science and media is of utmost importance for democratic citizenship and how people act in relation to risks and crises.

In Chapter 9, "Mediated science and issues of public knowledge and trust", Anna Maria Jönsson discusses public communication and images of science in Swedish news media. Based on a mixed-methods analysis of empirical data covering three decades, Jönsson argues that certain disciplines, mostly in the science, technology, engineering, and mathematics (STEM) disciplines, tend to gain more media coverage, are viewed as more trustworthy and are seen as generally important for the public good both by journalists and by audience members. As Jönsson notes, the level of trust in research and researchers within society is crucial for effectively addressing the challenges that societies face. While maintaining a critical perspective towards elites and public statements is valuable to a certain extent, trust in institutions and individuals remains essential for the functioning of democratic societies. It is important to protect research from being manipulated by political or ideological agendas, especially in the humanities, where trust in research is notably polarised.

Chapters 10 and 11 shift the focus to one particular field of science and engineering that is particularly loaded in its public image: nuclear energy. In Chapter 10, "Constructing trust through affective discipline: Finnish nuclear energy experts and the Fukushima Daiichi disaster", Anna Rantasila examines how mainstream news media and nuclear energy experts in Finland sought to construct and maintain trust in nuclear energy during the 2011 Fukushima Daiichi disaster. According to Rantasila, during a crisis situation where reliable information is scarce and the situation is in constant flux, a key element in constructing trust is management of public emotions. Based on her analysis, Rantasila argues that in addition to relaying the public with accurate knowledge, experts and journalists must be mindful of the emotional impact a crisis situation has on the public.

Following Rantasila's account of a specific nuclear incident's media coverage, Chapter 11, "Nuclear stories in the news media: Filtering and altering of expert views" by Gabor Sarlos analyses risk-related press releases by the International Atomic Energy Association (IAEA) and their impact on global news coverage of nuclear energy in 2023. According to Sarlos, the press releases by the IAEA provide a dual image. On one hand, press releases highlight substantial risk factors, for example, the escalation of military activities in the imminent vicinity of a nuclear power plant or the risk of not providing a sufficient amount of water for reactor cooling. On the other hand, the IAEA is quick to attenuate risk factors by putting in the focus its presence in the field, personal engagement, and technical expertise. The news media adds another layer to this dynamic. According to Sarlos, when the original source downplays the risk content of the message, news media might significantly counterbalance this by amplifying it.

In Chapter 12, "Journalist–source relations in Russian environmental journalism", Olga Dovbysh and Mika Perkiömäki explore the relationship between environmental journalists and news sources in Russia. Mutual influence in journalist/source relations and consequent journalistic decisions affects whose version of the world is being represented and reveal more general issues on social dominance and legitimacy. In such a controversial and politicised beat as climate change and related environmental problems, access to data and sources is playing a crucial role in the environmental agenda setting. Using in-depth interviews with environmental journalists working in Russian media, the chapter studies how sourcing is organised and what factors shape journalist/source relations in the context of a restricted media regime. According to Dovbysh and Perkiömäki, the agency and expertise of environmental journalists is limited by pressures from the owner and is rooted in the national journalistic culture's understanding of professionalism as following the "rules of the game" in sources selection. The war in Ukraine and growing pressure on media and civic society in Russia would negatively impact the representation alternative to the state opinions on climate and environmental problems.

References

Achterberg, P., de Koster, W., & Jeroen van der W. (2017). A science confidence gap: education, trust in scientific methods, and trust in scientific institutions in the United States, 2014. *Public Understanding of Science, 26*(6), 704–720. https://doi-org/10.1177/09636 62515617367

Agley, J. (2020). Assessing changes in US public trust in science amid the COVID-19 pandemic. *Public Health, 183*, 122–125. https://doi.org/10.1016/j.puhe.2020.05.004

Agustín, O.G., & Jerne, I. (2025). Green populism: Counterpublics and the formation of counterknowledge. In M.M. Roslyng, A. Rantasila, & A.M. Jönsson (Eds.), *Communicating science, climate change and the environment in hybrid media: Constructed facts, contested truths* (pp. 70–88). Routledge.

Alinejad, D., & Van Dijk, J. (2023) Climate communication: how researchers navigate between scientific truth and media publics. *Communication and the Public*, *8*(1), 29–44. https://doi.org/10.1177/20570473221138612

Bergström, A. & Oscarsson, H. (2105). Svensk forskningsopinion 2002–2014 (Swedish Research Opinion 2002–2014). In A. Bergström, B. Johansson, H. Oscarsson, & M. Oskarsson (Eds.), *Fragment*. Göteborg: SOM-institutet vid Göteborgs universitet.

Bromme, R., Mede, N.G., Thomm, E., Kremer, B., & Ziegler, R. (2022). An anchor in troubled times: trust in science before and within the COVID-19 pandemic. *PloS One*, *17*(2). https://doi.org/10.1371/journal.pone.0262823

Büscher, B. (2021). *The truth about nature: environmentalism in the era of post-truth politics and platform capitalism*. University of California Press.

Carlson, M. (2018) The Information politics of journalism in a post-truth age. *Journalism Studies*, *19*(13), 1879–1888. https://doi.org/10.1080/1461670X.2018.1494513

Chadwick, A. (2013) *The Hybrid media system: politics and power.* Oxford University Press.

Crasnow, S. & Intemann, K. (2020). Introduction. In S. Crasnow & K. Intemann (Eds.), (2020). *The Routledge handbook of feminist philosophy of science* (pp. 1–9). Taylor & Francis Group..

Cunningham-Burley, S. (2006). Public knowledge and public trust. Community *Genetics*, *9*(3), 204–210. https://doi.org/10.1159/000092658

Davis, S.R. & Horst, M. (2016). *Science communication: culture, identity and citizenship*. Springer.

De Cleen, B. (2018). Populism, exclusion, post-truth. Some conceptual caveats. *International Journal of Health Policy and Management, 7*(3), 268–271. https://doi.org/10.15171/ijhpm.2017.80

Demata, M. (2025). Fighting (for) truth? Alex Jones, the WHO and the legitimation of conspiracy discourse. In M.M. Roslyng, & A. Rantasila, & A.M. Jönsson (Eds.), *Communicating science, climate change and the environment in hybrid media: constructed facts, contested truths* (pp. 91–111). Routledge.

Demata, M., Zorzo, V., & Zottola, A. (2022). *Conspiracy theory discourses*. John Benjamin Publishing.

Eslen-Ziya, H. & Giorgi, A. (Eds.) (2022). *Populism and science in Europe*. Palgrave Macmillan.

Fawzi, N. (2018). Untrustworthy news and the media as "enemy of the people?" How a populist worldview shapes recipients' attitudes toward the media. *The International Journal of Press/Politics, 24*(2), 146–164. https://doi.org/10.1177/1940161218811981

Ferrara, E., Cresci, S. & Luceri, L. (2020). Misinformation, manipulation, and abuse on social media in the era of COVID-19. *Journal of Computational Social Science, 3*, 271–277, https://doi.org/10.1007/s42001-020-00094-5

Fraser, N. (1990). Rethinking the public sphere: A contribution to the critique of actually existing democracy. *Social Text, 25/26*, 56–80. https://doi.org/10.2307/466240

Fuchs, C. (2020). Everyday life and everyday communication in coronavirus capitalism. *TripleC: Communication, Capitalism & Critique, 18*(1), 375–399. https://doi.org/10.31269/triplec.v18i1.1167

Goldenberg, M.J. (2020). Rethinking Debates About Pediatric Vaccine Safety. A Feminist View. In S. Crasnow, & K. Intemann (Eds.), *The Routledge handbook of feminist philosophy of science* (pp. 423–434). Taylor & Francis Group.

Goldenberg, M.J., Adhikari, B. von Seidlein, L., Cheah, Y.P. & Larson, H.J. (2023). Vaccine mandates and public trust do not have to be antagonistic. *Nature Human Behavior* https://doi.org/10.1038/s41562-023-01720-8.

Grech, A. (2021). *Media, technology and education in a post-truth society: from fake news, datafication and mass surveillance to the death of trust*. Emerald Publishing.

Hatakka, N., & Herkman, J. (2022). Hegemonic meanings of populism: populism as a signifier in legacy dailies of six countries 2000–2018. *Media, Culture & Society, 44*(8), 1523–1540. https://doi.org./10.1177/01634437221104680

Hendriks, F., Kienhues, D., & Bromme, R. (2016). Trust in science and the science of trust. In B. Blöbaum (Ed.), *Trust and communication in a digitized world: Models and concepts of trust research* (pp. 143–159). Springer International Publishing/Springer Nature. https://doi.org/10.1007/978-3-319-28059-2_8

Holt, K. (2018) Alternative media and the notion of anti-systemness: towards an analytical framework. *Media and Communication, 6*(4), 49–57. https://doi.org/10.17645/mac.v6i4.1467

Jasanoff, S. (2004). The idiom of co-production. In S. Jasanoff (Ed.), *States of knowledge: the co-production of science and social order* (pp. 1–12). Routledge.

Jönsson, A. M. (2020). Polariserat forskningsförtroende – utbildningsnivå och partisympatier vattendelare (Polarized trust in research – educational level and party affiliation as dividing lines). In U. Andersson, A. Carlander, & P. Öhberg (red) *Regntunga skyar*. Göteborg: SOM-institutet vid Göteborgs universitet.

Jungherr, A., Posegga, O., & An, J. (2019). Discursive power in contemporary media systems: a comparative framework. *The International Journal of Press/Politics*, *24*(4), 404–425. https://doi.org/10.1177/1940161219841543

Kuula, A. (2006) *Tutkimusetiikka. Aineistojen hankinta, käyttö ja säilytys*. [Research ethics. Collection, use and storage of empirical data]. Vastapaino.

Larsen, G. & Roslyng, M.M. (2025). Knowledge and counter-knowledge: The construction of facts in vaccination debates. In M.M. Roslyng, A. Rantasila, & A.M. Jönsson (Eds.), *Communicating science, climate change and the environment in hybrid media: constructed facts, contested truths* (pp. 112–130). Routledge.

Lewis, M., Govender, E., Holland, K. (Eds.) (2021) *Communicating Covid-19: interdisciplinary perspectives*. Springer International Publishing

Luengo, M. & García-Marín, D. (2020). The performance of truth: Politicians, fact-checking journalism, and the struggle to tackle COVID-19 misinformation. *American Journal of Cultural Sociology*, *8*(3), 405–427. https://doi.org/10.1057/s41290-020-00115-w

Lynch, M. (2016). Social constructivism in science and technology studies. *Human Studies, 39*(1), 101–112. https://doi.org/10.1007/s10746-016-9385-5

Mäkinen, M. (2025). Citizen activists or pandemic deniers? Alternative voices in the Finnish journalistic media during the COVID-19 pandemic. In M.M. Roslyng, A. Rantasila, A.M. Jönsson (Eds.), *Communicating science, climate change and the environment in hybrid media: constructed facts, contested truths* (pp. 131–150). Routledge.

Mede, N.G., & Schäfer, M.S. (2020). Science related populism: conceptualizing populist demands towards science. *Public Understandings of Science*, *29*(5), 473–491. https://doi.org./10.1177/0963662520924259

Mouffe, C. (1999). Deliberative democracy or agonistic pluralism? *Social Research*, *66*(3), 745–758. www.jstor.org/stable/40971349

Mouffe, C. (2005). *On the political*. Routledge.

Oreskes, N. (2019). *Why trust science?* Princeton University Press.

Pinto, J., Gutsche, R.E., & Prado, P. (Eds.). (2019). *Climate change, media & culture: critical issues in global environmental communication*. Emerald Publishing.

Portelli, J.P. & Oladi, S. (2021). Post-truth society: towards a dialogical understanding og truth. In A. Grech (Ed.), *Media, technology and education in a post-truth society* (pp. 11–28). Emerald Publishing Limited.

Rantasila, A. (2020) *Circulating emotions, sticky feelings: affective dynamics of the Fukushima Daiichi nuclear disaster in a hybrid media environment*. Doctoral Dissertation, Tampere University. https://urn.fi/URN:ISBN:978-952-03-1764-5

Rantasila, A. (2025) Constructing trust through affective discipline: Finnish nuclear energy experts and the Fukushima Daiichi disaster. In M.M. Roslyng, A. Rantasila, & A.M. Jönsson (Eds.), *Communicating science, climate change and the environment in hybrid media: constructed facts, contested truths* (pp. 176–193). Routledge.

Rolin, K. (2020). Situated knowledge and objectivity. In S. Crasnow, & K. Intemann, (Eds.), *The Routledge handbook of feminist philosophy of science* (pp. 216–224). Taylor & Francis Group.

Roslyng, M.M. (2022). Between populism and popular citizenship in science conflicts. In H. Eslen-Ziya, & A. Giorgi (Eds.), *Populism and science in Europe* (pp. 207–229). Palgrave Macmillan.

Roslyng, M.M. & Larsen, G. (2021). The struggle between the power of language and the language of power. Pro- and anti-vaccination memes and the discursive construction of knowledge. *Journal of Language and Politics 20*(6), 894–912. https://doi.org/10.1075/jlp.19101.ros

Schäfer, M.S. (2009). From public understanding to public engagement: an empirical assessment of changes in science coverage. *Science Communication*, *30*(4), 475–505. https://doi.org/10.1177/1075547008326943

Schatto-Eckrodt, T., Clever, L., & Frischlich, L. (2024). The seed of doubt: examining the role of alternative social and news media for the birth of a conspiracy theory. *Social Science Computer Review*, *42*(5), 1160–1180. https://doi.org./10.1177/0894439324 1246281

Sorensen, L. & Krämer, B. (2024). The shift to authenticity: a framework for the analysis of political truth claims. *Communication Theory*, *34*(3), 118–129. https://doi.org/10.1093/ct/qtae013

Speed, E. & Mannion, R. (2017). The rise of post-truth populism in pluralist liberal democracies: challenges for health policy. *International Journal of Policy and Management*, *6*(5), 249–251. www.ijhpm.com/article_3322.html

Suiter, J. (2016). Post-truth politics. Political *Insight*, *7*(3). https://doi.org/10.1177/20419 05816680417

Sulik, J., Deroy, O., Dezecache, G., Newson, M., Zhao, Y., El Zein, M., & Tunçgenç, B. (2021). Facing the pandemic with trust in science. *Humanities and Social Sciences Communications*, *8*(1), 1–10. https://doi.org/10.1057/s41599-021-00982-9

Tuomola, S. (2023). *Hyvä, katala vastajulkisuus: Populistisen vastamedian julkisen puhuttelun tavat maahanmuutto- ja pakolaiskeskustelussa*. [Good and bad counterpublics: modes of public address in populist countermedia's immigration and refugee debates]. Doctoral dissertation, Doctoral Programme in Communication, Media and Theatre, Faculty of Information Technology and Communication Sciences, Tampere University. https://urn.fi/URN:ISBN:978-952-03-2724-8

Valdivia, A.N. (2013). Media studies. The interdiscipline of the present and the future. In J. Nerone (Ed.), *The international encyclopedia of media studies, volume I: media history and the foundations of media studies* (pp. 1–18). Blackwell.

Van Dijk, J. & Alinejad, D. (2020) Social media and trust in scientific expertise: debating the covid-19 pandemic in The Netherlands. *Social Media + Society*, 1–11, https://doi.org/10.1177/2056305120981057

Vécsey, V. (2025). Web of denial. Climate change denial discourse on Instagram. In M.M. Roslyng, A. Rantasila, & A.M. Jönsson (Eds.), *Communicating science, climate change and the environment in hybrid media: constructed facts, contested truths* (pp. 35–56). Routledge.

Vraga, E.K. & Bode, L. (2020). Defining misinformation and understanding its bounded nature: using expertise and evidence for describing misinformation. *Political Communication, 37*(1), 136–144. https://doi.org/10.1080/10584609.2020.1716500

Weignart, P. (2022). Trust or attention. Medialization of science revisited. Public Understanding of Science, *31*(3), 288–296. https://doi.org/10.1177/09636625211070888

Ylä-Anttila, T. 2018. Populist knowledge: 'post-truth' repertoires of contesting epistemic authorities. *European Journal of Culture and Political Sociology, 5*(4), 356–388. https://doi.org/10.1080/23254823.2017.1414620

Part 1

Environmental and climate truths in media

2 The scientification of risks and the risks of scientification

Insights from the coverage of artificial turf pitches as microplastic pollutants in Sweden

Ernesto Abalo

Introduction

"I want you to listen to the scientists", Swedish environmental activist Greta Thunberg told the US Congress in 2019 (NBC News, 2019) in a quote that would go viral on social media, establishing a slogan for environmentalists seeking political action against the never ending, even escalating, climate crisis. Examining her words in depth, the quote is not only illustrative of how science is often treated in public discourse about environmental issues; it also unveils the hope that sometimes is put on science. By calling for attention on scientists, Thunberg makes science the realm in which the truth about the climate crisis lies, and scientists the ones carrying this truth. Thunberg could instead have begged Congress to listen to environmentalists or to listen to the people of the Global South, also actors carrying truths about the climate crisis – however, those actors are without the authority, credibility, and knowledge legitimacy of science. Moreover, by using a definite article, "the scientists", researchers across disciplines are grouped together as having one voice and the same conclusions. This perception of unanimity has to do with the scientific consensus that has been constructed around the climate issue, both by scholars themselves, as in the Intergovernmental Panel on Climate Change (IPCC) reports (see IPCC, 2023), but also by the news media (Olausson, 2009). Moreover, referring to "the scientists" also serves political purposes, as it connects Thunberg's arguments with scientific knowledge. In this sense, "the scientists" in Thunberg's call become a sort of Messiah-like agent, holding the key not only to truth but to salvation.

This phrasing of Thunberg is an example of a discursive strategy that will be treated in depth in this chapter, namely that of scientification (Carvalho, 2005). Scientification, in gross terms, is the discursive framing of an issue, in this case an environmental one, as a scientific matter. It is when we brand a topic as something to be understood by the rules of science. Scientification can be used in different forms of public discourse. As in the example of Thunberg, it can be used in the sphere of politics, as artillery in a debate. But as the same example also shows, scientification can be captured and distributed by the media and become part of media discourse.

DOI: 10.4324/9781003479550-3

Although operating under different logics (Olausson & Berglez, 2014; Schäfer, 2017), science and the media play important roles for each other. The media is a window for researchers to reach out to the public; for citizens, the media is an important source of information about science, and for the media, science is not only a topic of news itself but also a resource for facticity and credibility (see Jönsson, 2025; Rantasila, 2025). In the midst of the fear of a "post-truth" era (Farkas & Schou, 2020; McIntyre, 2018), where "fake news" and "alternative facts" circulate while objectivity and accuracy are relativised, science and *the facts* it provides, stands on the other end of the spectrum and can be seen as the antidote. By relying on science and scientific experts, media outlets can thus safe-mark themselves in a situation where the assessment of information quality and accuracy has become increasingly important (see Dovbysh & Perkiömäki, 2025).

At the same time, *the facts* are somewhat sensitive to the uncertainty that lies at the core of science, which provides a challenge for the media when science-centred topics, such as environmental risks, are transformed into media content. Risk revolves around the "probabilities and consequences of adverse events" (Slovic, 2006, p. 690). However, as Slovic (1999, p. 690) points out, risks do "not exist 'out there', independent of our minds and cultures, waiting to be measured". The concept itself has been invented to help us "understand and cope with the dangers and uncertainties of life" (Slovic, 1999, p. 690), making risk assessment prone to subjectivity. Moreover, risks are not only assessed but also constructed and communicated, making the news media an important co-constructor of risks (Höijer, 2009). In the case of environmental risks, where science is a central realm for risk assessment, the recontextualisation of risk, thus its transformation from science to media discourse, may lead to the renegotiation of scientific assessments. As scholars have shown, journalistic routines, and what is commonly called media logic (Altheide & Snow, 1979), can serve for the contestation of science-based claims in the media (Boykoff & Boykoff, 2004; Lidskog & Olausson, 2013; Peters & Dunwoody, 2016; Schäfer, 2009). To some extent, this type of contestation has to do with the media's centration on problems and conflict (see Altheide, 1997; Karidi, 2018), which in the case of reporting on risks, can have the effect of highlighting risks that have meagre scientific support (Völker et al., 2020). Therefore, there is a necessity of understanding how the media handles science in the construction of risks, such as those risks related to the environment.

The aim of this chapter is to explore how scientification is used in media discourse in order to enhance understanding of how the embeddedness of science in specific discursive strategies in news media shapes the reporting around environmental risks. By looking at the strategy of scientification in detail, it is possible to obtain a better understanding of the strengths and weaknesses of the ways in which science is represented in the media. Such an approach also enables one to understand how science, through the use of specific discursive means, becomes a key component in the construction of different arguments around environmental risks. This chapter centres on the news discourse on a specific environmental risk, that of artificial turf pitches as microplastic pollutants in Swedish aquatic environments, and draws on a critical discourse analysis of Swedish news media's coverage of the case (see Abalo, 2019a), commissioned by Svensk däckåtervinning AB (Swedish Tyre Recycling, SDAB). A detailed look at the use of scientification in this specific case unveils the dynamic and complex relationship between science, environmental risks, and media discourse.

In the following section, the concept of scientification is presented in more detail; that is followed by an account on the risks of microplastics and artificial turf. Next, the data and analysis that this chapter relies on is presented, which is followed by examples of the use of scientification in the media coverage of artificial turf. Finally, some concluding remarks are presented.

Scientification as a discursive strategy in the media

On one level, scientification can be understood as the organisation and professionalisation of knowledge about something. Scientificaton in that sense, explains the emergence of an academic discipline (see Jung, 2015; von Stuckrad, 2014). Moreover, scientification can also be understood as "the social processes whereby more and more of everyday life has come into the scientific domain and is subject to scientific influence and supervision" (Yamaguchi, 2016, p. 69). According to such understanding of the concept, scientification refers to the principles of science moving from academia to the everyday lives of people.

On another level, related to these understandings and yet different, scientification can be perceived as a discursive strategy. This understanding is rooted in the tradition of critical discourse analysis (CDA), an academic discipline involving theory and methods that aims to understand the relationship between language and power (Fairclough, 1995). Discourse is here understood as language use and as a form of social practice that is dialectically connected to different social structures (Fairclough & Wodak, 1997). According to this understanding, discourse has ideological implications through reinforcing or challenging power relations. CDA offers a plethora of different categories through which to analyse language use; one of these is discursive strategy. A discursive strategy is the way an actor more or less consciously accommodates language to reach a specific goal (Carvalho, 2008). As Carvalho (2005, p. 3) explains, discursive strategy is "a transformative discursive move involving the semantic re-definition of an object (or actor)". There are different discursive strategies, including, for example legitimation (justifying actions) and politicisation (attributing a political nature to something) (Carvalho, 2008).

As a discursive strategy, scientification is here understood as the attribution of a scientific nature to an issue. In contrast to other discursive strategies, the literature on scientification is limited. Carvalho (2005) identifies scientification as a specific discursive strategy in her study on the British press coverage of climate change. There, Carvalho notes that early in the treatment of the question of climate change, in 1989, Margret Thatcher is quoted as asking for scientific evidence before anything could be done about the climate question. Carvalho categorises this discursive approach as an example of scientification and further notes that Thatcher's specific use of the strategy is problematic, as it pleads for not acting on a precautionary basis. Scientific uncertainty was thus used as an argument for not interfering on global warming. Carvalho (2005, p. 8) labels the use of scientification by social actors as "analytical", meaning that these actors are using scientification to analyse what something, climate change in this case, is about.

Inspired by Carvalho's use of scientification, I have found the category useful when analysing discursive strategies in the media discourse about artificial turf (Abalo, 2019a) and medical cannabis (Abalo, 2021). My usage is a bit more flexible than Carvalho's, as I do not solely consider scientification as an analytical strategy. Furthermore, Carvalho's (2005) use of scientification is centred on utterances by specific actors quoted in the media and on instances of the journalistic voice. Although such utterances are indeed key "places" where scientification is expressed, we also need to look beyond what is manifested in text. We must look beyond the text because the attribution of a scientific nature to an issue is not solely realised when social actors say something science oriented about an issue, it is also realised by newsroom decisions, or production strategies, that pave the way for the attribution of a scientific nature to an issue (see Allan et al., 2010).

One such production strategy could be to frame a story around science beginning from the headline or lead paragraph, another would be to make scientists, researchers or other types of science-bound experts key sources in the media (see Allan et al., 2010; Leidecker-Sandmann et al., 2022; Lidskog & Olausson., 2013). These actions seldom happen by chance but are the outcomes of newsroom decisions. In my own research on journalism about cannabis, for example, I have noted that it is common for doctors or psychologists to be interviewed about the effects of the drug or its medical potential, which leads to quotations that could be categorised as scientification (Abalo, 2021). In such cases, scientification is enabled by how the newsrooms organise their production of their news reports; this practice is commonplace, as science-oriented experts have certain authority and by citing them journalism becomes credible. This production strategy is, of course, visible on several environmental questions, as scientific knowledge is important for explaining and understanding these questions.

Scientification is not only found in quotes from science-related sources, it is also, as Carvalho's example of Thatcher shows, visible in the language use of other actors. In my research on the media discourse on medical cannabis, I have found that many sources refer to science, but that scientific *precision* on the topic differs (Abalo, 2021). This has to do with what, within CDA, is called *recontextualisation* (Fairclough, 1995). Recontextualisation is when text, a sign or meaning from one context is placed in a new context and is adapted to the conventions of the new context. Science, in an academic context, is subjected to a set of rules through which it is practiced. In a news media context, however, many of the conventions that define science in an academic context are inapplicable. Instead, science appears in other forms: in interviews with researchers, in people's arguments, in vague references to "science" itself. In other words, scientification in news media discourse means that science is always recontextualised and represented.

In my previous work, I have differentiated between strong science discourse and weak science discourse (Abalo, 2021). Strong science discourse (or strong scientification) is when a statement is linked to specific actors in the scientific community or to specific studies, while weak science discourse (or weak scientification) is when a statement is linked to science much more loosely or vaguely. I find this distinction important to make, as many actors refer to science in order to give their perspective certain weight and legitimacy, but not all these references carry the same weight or have the same precision. A researcher being quoted speaking from her or his professional knowledge (strong scientification) is something apart from a layperson claiming that something "is supported by science" without pointing to specific studies (weak scientification); these two different types of scientification can be used with specific purposes by the news media. Two scientists with conflicting science-rooted opinions can serve to construct conflict, which is a way for the media to make a story relevant and at the same time remain neutral. A layperson claiming something with vague reference to science and then being countered by a scientist pointing to statistics or specific studies can also be a way for the media to create conflict but, at the same time, provide more weight to the argument of the scientist because of the scientist's position and the stronger foundation of the scientist's argumentation. The use of strong and weak scientification points to the fact that science, when recontextualised and represented in this way, somewhat loses the authority to decide what holds as scientific, while the media and quoted experts obtain the power to define the scientific nature of an issue.

Before investigating the use of scientification in the Swedish media, I provide an account of the risks of microplastics and artificial turf in Sweden. This is done in order to provide contextual knowledge, which is important for understanding the use of scientification in the media discourse around this example.

Microplastic risks and the media

Although the topic of microplastics has a long history in research in ecotoxicology and marine biology, the risks of microplastic pollution in aquatic environments has become a hot public environmental topic in several European countries in recent years (see Burton & Cervi, 2019). Microplastics, or solid plastic particles no bigger than 5 mm in diameter, can either be newly produced polymer-containing particles (sometimes called primary microplastics) or disintegrated particles from larger plastic items (or secondary microplastics) (ECHA, n.d.; Lambert & Wagner, 2018). According to the European Chemicals Agency (ECHA), about 145,000 tonnes of microplastics are used in Europe each year, of which 42,000 tonnes are released into the environment (ECHA, n.d.). In 2019, ECHA proposed a restriction on intentional uses of microplastics (ECHA, 2019), and there has been national bans of so-called microbeads (smaller microplastic particles) in cosmetic products (ECHA, n.d.).

In 2017, ecotoxicology professor G. Allen Burton (2017) wrote an opinion piece in *Environmental Science and Technology* in which he argues that even the highest concentrations of microplastics in aquatic environments are far too low to cause any big risk, and that one needs to distinguish between real concentrations of microplastics in aquatic environments and concentrations found in laboratory studies. Burton also criticises research papers claiming that microplastics are an environmental threat being "rapidly picked up by the news media" and thereby wrongfully influencing decision-makers (Burton, 2017: 13515). Burton's commentary has drawn both sympathy and disproof, as can be seen in Backhaus and Wagner's (2020) debate on the magnitude of the microplastic risk. A research review by Science Advice for Policy by European Academies (SAPEA, 2019) does not find microplastics to pose a widespread risk to the environment or to humans. However, the report points to limitations in how levels of microplastics are measured, as well as to eventual risks if pollution continues at the same rate.

Burton's argument that the media picks up alarmist notions on microplastics is somewhat echoed in media research. When comparing how the issue of microplastics is framed in peer-reviewed journals and online media, Völker et al. (2020) found some interesting differences. While scientific journals frame environmental risks as hypothetical or as surrounded by uncertainty, the outcomes of microplastic risks are constructed harmful in over 90% of analysed media articles. In other words, "the framing of the media articles implies that harmful consequences are highly probable" (p. 7). According to Völker et al., this contrast has to do with differences in how risks are understood, or "conflicting risk concepts" (p. 9). The conflicting risk concepts are that while science treats risk as "the probability of a negative outcome", in public discourse, risk is instead perceived as "the uncertainty of a negative outcome itself" (p. 9). Therefore, uncertainty, which is in fact an integral part of science, tends to feed risk constructions in the media. While much is unknown about the environmental risks of microplastics, they tend to be boosted in the media.

Relatedly, the media's treatment of plastic pollution (in a broad sense) seems to have repercussions on other areas of public discourse. Males and Van Aelst (2020) studied the agenda-setting role of the documentary *The Blue Planet* in establishing the issue of plastic pollution in British public discourse. According to the researchers, plastic pollution became more frequently mentioned both in the media and in politics after the documentary was aired in 2017. This suggests that popular media content, such as *The Blue Planet*, may impact how environmental risks, such as plastic pollution, become relevant for other media and may also serve to shape political discourse on the topic. With Völker et al.'s (2020, p. 9) "conflicting risk concepts" in mind, it is reasonable to believe that there is a possibility that policies on plastics and microplastics are based on risk perceptions that do not align with what is considered a risk from a scientific point of view. On the positive side, that could lead to precautionary measures. On the negative side, it could lead to misdirected policies.

The case of artificial turf in Sweden

In Sweden, where this study is based, the risks around microplastic pollution became a political issue in 2015. That year, the Swedish Environment Protection Agency (SEPA) was commissioned by the government to study the sources of microplastic pollution in marine environments and to find solutions. The independent IVL Swedish Environmental Research Institute (IVL), which executed the survey for the SEPA, issued a first report in 2016 (Magnusson et al., 2016) that was updated a year later (Magnusson et al., 2017). According to IVL, the top two sources of microplastic pollution in the sea are tyre wear from road traffic (13,000 tonnes per year) and artificial turf pitches used for playing football (2,300–3,900 tonnes per year). In the case of artificial turf pitches, it is not primarily the straws that are considered an environmental risk but the gum granules placed on top of the pitches to increase playing capabilities. These granules are frequently produced from worn out tyres, and this production is, at least from the point of view of the producers, conceived of as a way to recycle that type of tyre waste. The IVL figures sparked a public debate in Sweden about the environmental impact of artificial turf pitches. As football is the most popular sport in Sweden, and artificial turf pitches have become numerous and popular because they allow year-round activity in a country where football would otherwise be limited to the summer season, IVL's figures sparked some controversy and put artificial turf pitches under the microscope. In the following years, the SEPA proposed a notification obligation for those wishing to install new pitches and the creation of a finances costumer group to provide support for municipalities planning to build new pitches (see BEKOGR, 2023).

However, IVL were open about the uncertainty that surrounded the initial figures on microplastic pollution. In the report, IVL states that for "several sources suspected to contribute with large amounts of microplastics to the sea, data is so scarce that no estimations on emissions could be done" and points out that the ranking data "suffers from a large degree of uncertainty" (Magnusson et al., 2017, p. 6). In the case of artificial turf pitches, the authors state that they do not know how much of the spread of artificial turf microplastics reaches the sea (Magnusson et al., 2017), which is due to the fact that the institute did not measure this spread. Instead, what IVL measured was the amount of gum granules that were refilled, and figures for this were obtained through personal communication with providers and municipalities. The amount of granule that was refilled was assumed to be the same amount that was spread from the pitches. A later report from IVL, using a more sophisticated approach to measuring the spread of granules, estimated that the loss was significantly – about ten times – lower (Krång et al., 2019).

The environmental risks of artificial turf pitches, as proposed by IVL, drew attention from the Swedish news media at both the national and local levels. A content analysis of Swedish media shows that it was in 2016, the same year that IVL published its first report, that artificial turf started to be associated with plastic and gum pollution (Abalo, 2019b). The same study also showed that the environmental gains of artificial turf, such as tyre recycling, were not normally part of the coverage, while only about every third article mentioned uncertainty in relation to the environmental impacts of artificial turf.

Looking more closely at the discursive level of the media coverage, it is apparent that the discursive construction of artificial turf has been very much risk oriented (Abalo, 2019a). A Swedish news consumer might encounter headlines such as "The football pitch could be a ticking environmental bomb" (*Svt*, 21 January,

2017), "Thousands of tons of gum disappear from artificial turf pitches each year" (*Svt*, 17 January, 2017), and "Artificial turf spreads microplastics" (*Göteborgs-Posten*, 21 January, 2017), pointing to a risk-oriented framing.[1] The fact that the figures on artificial turf came from a research institute commissioned by the SEPA likely made newsrooms consider the reports from the institute to be reliable, despite the uncertainty surrounding the figures. Very much in line with the arguments of Völker et al. (2020), uncertainty did not hinder the construction of risk in the artificial turf case. This risk construction relates to how journalists and experts let uncertainty be moulded by the logics of the media in the production of news on the artificial turf issue (Abalo & Olausson, 2023).

Grasping scientification in the case of artificial turf – A note on method

The data used in this chapter derive from a larger study on the discourse on artificial turf in Swedish news media (Abalo, 2019a). That study was part of a research project commissioned and funded by SDAB.[2] The original sample consisted of 160 articles from nine news outlets. These outlets are the newspapers *Dagens Nyheter* (*DN*), *Svenska Dagbladet* (*SvD*), *Aftonbladet* (*AB*), *Göteborgs-Posten* (*GP*), *Sydsvenska Dagbladet* (*Sydsvenskan*), *Nerikes Allehanda* (*NA*), *Norrköpings Tidningar* (*NT*), *Norrbottens Kuriren* (*NK*), as well as the website (two local desks) of the Swedish public service television network *Sveriges Television* (*Svt*). Newspaper materials were collected through searches on Retriever Mediearkivet, and *Svt* materials were collected through Google searches. Articles published between January 1, 2016, and August 20, 2019, that discursively constructed artificial turf within an environmental context were included. The sample contains news articles, features, opinion pieces, briefs and blurbs.

The CDA conducted in the larger study was based on Carvalho's (2008) model for discourse analysis of journalistic texts. This involves analysis of the layout and structural organisation of an item (including graphic items), the objects (or themes) present, the social actors appearing in the text, language and rhetoric, the discursive strategies used, and ideological positions that can be induced from the analysis of the other categories. Furthermore, a synchronic analysis was made to distinguish patterns in the different texts.

This chapter is rooted in the sample from the broader study but, analytically, only focused on discursive strategies and, more specifically, on the strategy of scientification. This has meant reanalysing some of the articles in order to make the analysis fit the purpose of the chapter. The analysis of scientification began with the identification of items that refer to science, researchers, studies, or any other lexical choices that place discourse on artificial turf or microplastics in a scientific context. It is in such items that one finds the discursive strategy of scientification. The next step has been to examine the ways in which scientification is used, thus how artificial turf and microplastics are discursively connected to science, if that connection is a case of strong or weak scientification, how such constructions fit to the general framing of an item, and what discursive purposes such connection fills. Moreover, the analysis has sought for discursive patterns in the use of scientification.

Examples of scientficiation in relation to artificial turf and microplastics

In this section, I provide examples of three different ways in which scientification appears in the reporting on microplastics and artificial turf: scientification is used to attach weight and certainty to claims, to highlight uncertainty and as a means for fallacy.

Scientification for weight and certainty

The use of scientification in the coverage of artificial turf unveils the difficulties that the recontextualisation of science causes for journalism. What counts as scientific research and what does not is a difficult concept to handle for newsrooms, which results in some ambiguity. For some newsrooms, IVL's report on microplastics counted as scientific research, which provided weight to the report and to the environmental problem it was addressing.

> **Research: Artificial turf pitches dangerous for the environment**
>
> Microplastics pollute our oceans. And artificial turf pitches are one of the villains when it comes to littering, new research from the Swedish environmental institute shows.
>
> (*Sydsvenskan*, 9 March 2016)

> When researchers studied potential sources of microplastics in the sea last year, artificial turf pitches came in second place. The maintenance of artificial turf pitches must be improved in order to decrease loss, claims the environmental engineer [Name].
>
> (*Svt*, 17 January 2017)

Both of these excerpts refer to IVL's report on microplastics and are attempts to construct strong scientification by the newsrooms. By making reference to "research" and "researchers", the newsrooms are providing relevance to the case of artificial turf as an environmental risk.[3] Whether the report is really scientific or not seems to be less relevant for the newsrooms, which shows how flexible the treatment of science can be for the media. The use of science-related terms, such as "research" or "researchers" in the headline or the lead paragraph, as in the excerpts above, also indicates that scientification is part of the main framing of these articles. In turn, this suggests that scientification in these cases is very much the outcome of newsroom decisions, as it was decided to make science part of the selling point of the news pieces.

In the cases above, scientification is used to highlight artificial turf as a risk, which is, however, not always the case. In a rare example of a news article that criticises IVL's estimations on microplastics, a newspaper ran the headline "Research shows: Artificial turf is not that environmentally dangerous" (*Göteborgs-Posten*, 4 June, 2019), some years after the publication of the first IVL report. In this piece, the newspaper also chose to turn to science to frame the article and to establish a fact – or rather, to overturn an established fact. As the examples above demonstrate, scientification here also serves to construct certainty.

There are also examples of how scientification, mainly weak scientification, is used to provide weight to arguments about artificial turf. This type of use can be seen both in attempts to strengthen the argument of artificial turf as an environmental risk and to discursively mitigate artificial turf's risk. In either case, scientification is used to construct certainty in some way. A letter to the editor criticising the use of granules produced from worn out tyres argues that "Several independent researchers have warned about the use of this material" (*Göteborgs-Posten*, 5 September, 2017), without specifying who these researchers are or where and when the studies were published. In another article, an industry representative is quoted as saying that "We know that microplastics that are spread to lakes and the sea can have severe consequences, but not one of all the research studies has concluded that recycled tyre material has been spread there" (*Göteborgs-Posten*, 23 September, 2017). Here, the actor is a bit more specific, referring to "not one of all the research studies", but still leaves many things unanswered. The reader is not told what sample of studies the industry representative is referring to, or if any of those studies even surveyed the presence of granules from tyre material. As the reporter does not ask, at least from what is visible in the article, these questions remain unanswered. The "weakness" of this form of scientification lies in the vague way science is referred to, which in turn illuminates how science can become hollow when recontextualised to media discourse.

Scientification to highlight uncertainty

At the same time that scientification is used to construct certainty around the environmental risks of artificial turf pitches, as seen above, there are also examples of scientification that serves to highlight scientific uncertainty. Such examples, which often appear in quotations from social actors, highlight the limits of scientific knowledge about this environmental risk, inevitably pointing to the need for more research on the matter. For example, when explaining the use of gum granules on football pitches, an official from the Swedish football association is quoted as saying that "Something that I find important is to consider the research that is available, which is not a lot at the moment" (*Svt*, 19 January, 2017). By pointing to scientific uncertainty, which is done by highlighting that there is "not a lot" available research, this official is to some extent justifying the use of granules and disregarding precautionary action. In a response to a debate article criticising IVL's first report, a representative from IVL says that

> It is also very important to note that we are not saying that all these microplastics reach the sea, or even the closest watercourse. Much research remains to be done before we can pronounce ourselves clear on how much microplastic there is in the sea in total, where it comes from and what effect it has on the ecosystem.
> (*Sydsvenskan*, 20 December 2016)

Scientification here constructs uncertainty by pointing out that much "research remains to be done" before one can pronounce oneself on the amount of granules spread to aquatic environments. For the IVL representative, scientification here serves to disconnect IVL's report from some truth claims that have been made (by others) on artificial turf pitches' environmental harms. This move is interesting, as some media outlets have used IVL's figures as (scientific) evidence for the environmental risks of artificial turf.

The existence of constructions of uncertainty around the environmental impacts of artificial turf pitches must be seen within the overall framing of the issue in the media. There are examples that suggest that uncertainty is relativised when appearing within a strong framing of risk, as uncertainty, on the one hand, is acknowledged by being central in the use of scientification by specific social actors, while, on the other hand, is somewhat neglected in the overall framing of a story. For example, a news article titled "So will plastic emissions from artificial turf pitches decrease" (*Dagens Nyheter*, 17 December, 2017), already constructs environmental hazards in relation to artificial turf in the title, as the title presupposes that there are plastic emissions from pitches. This risk framing is strengthened in the lead paragraph, which states that "Artificial turf pitches emit thousands of tonnes of microplastics each year". The journalistic voice in the body text then refers to IVL's figures (credited to the SEPA) and states that the assessed amount of microplastics is spread in the environment. The first actor quoted is a politician, who begins by saying that "The emissions are not reasonable and cannot continue. Now that we know this, we need to act". In the last part of the article, an ecotoxicologist is quoted as saying that one needs to do a "lifecycle analysis" of the pitches to make the right decisions about them, and she asks "What environmental impacts do different types of granules have, how big of emissions do they cause, and what environmental consequences do they have from production to recycling[?]" The example of this news article shows how scientific uncertainty, which is what the quoted researcher is pointing to, is pushed back, while the main focus is placed on established facts about the environmental risks of artificial turf. In this sense, the scientification is in disharmony with the overall framing of the news piece.

Scientification as fallacy

The coverage of artificial turf and microplastics also contains examples where scientific findings are used to substantiate far-fetched claims, and to some extent, the previous examples have already shown that. However, I will here present an example that I find a bit more radical, in which scientification serves as the basis for fallacy. Fallacy is here understood as faulty reasoning, and the example I provide resembles what Musi and Reed (2022) call informal fallacy, which is faulty reasoning that could appear reasonable in a specific context.

An internet article titled "Microplastics found in Växjö's bathing lakes: 'Surprisingly high concentrations'" (*Svt*, 7 May, 2019) reports that different microplastics are found in new samples from the lake Växjösjön. The lead paragraph ends by stating that "One source of these could be the so-called gum granules that, over several years, have been spread on the municipalities' plastic-made football pitches", which constructs artificial turf pitches as a potential source. The journalistic voice goes on to explain that the gum granules come from tyres, that tyres contain "dangerous chemicals and plastic", that the granules are not harmful if they stay on the pitch, but that they can be dangerous for fish and other animals if they are consumed. The journalistic voice also states that the SEPA has identified artificial turf pitches as a significant source of microplastic emissions in Sweden, and that there are large knowledge gaps about the environmental impact of these pitches. The article then quotes two researchers:

> [Researcher 1] is a chemist at [University 1]. She has found different microplastics in new samples from Växjösjön, where one of the sources could actually be gum granules:
>
> "It shows surprisingly high concentrations of microplastics. During a course at [University 1] we have made a small experimental study. We raised our eyebrows when we saw our results, compared to what has been found previously in Swedish lakes".
>
> **The development worries**
>
> [Researcher 2] is a PhD student at [University 2] and she has also seen this in studies.
>
> "It has been shown to leak out in the environment. We can find it in our environmental samples. Both this grass that the pitch is made of and also these old tyre particles", she says.
>
> (*Svt*, 7 May 2019)

Later in the *Svt* article, a representative from the municipality is quoted in a statement that ends the article as saying that he is "worried that there is so much microplastic [found] in our tests. One can also wonder how one got the idea to have ground down tyres on the football pitches". As the article centres on the environmental hazards of artificial turf pitches, at first glance, it could seem that the quotations from the researchers prove the claim that there are gum granules from artificial turf pitches in the lakes of the city of Växjö. However, looking closely at the quotations, that is not the case. When presenting the findings of Researcher 1, the journalistic voice speculates that one of the sources of the microplastics that were found "could be" gum granules. Thus, Researcher 1 does not specify the type of microplastics that were found. Researcher 2, in contrast, does point to the presence of materials coming from artificial turf pitches. However, it remains unclear if the particles found by Researcher 2 are from Växjösjön or even from the city Växjö (University 1 and 2 are located in different Swedish cities). What is stated is that the researcher has found such particles in "studies"; she is not necessarily pointing to the same study as Researcher 1. The statements of the two researchers therefore do not really substantiate the claim that microplastics from artificial turf pitches are found in Växjö. In this discursive context, where artificial turf, in public discourse, has become associated with environmental problems, and where the framing of this specific article very much underscores this association, this article uses scientification to underscore the environmental risk. Such a construction is impactful. The day after this article ran, another newspaper ran a brief on the same story, stating that unusually high levels of microplastics were found in Växjösjön, and that they are "believed to come from gum granule that has been used for several years to top dress the municipality's football pitches of artificial turf, reports *Svt Nyheter Småland*" (*Aftonbladet*, 8 May, 2019).

Concluding remarks

One thing that we can learn from the coverage of artificial turf and microplastics in the Swedish news media is that scientification is rather open-ended. In its ambition to provide certainty and *facts*, the media may turn to science, or what it believes is science, to attach credibility to the reporting on a specific issue. Scientification can then discursively appear in different ways: in the headlines and lead paragraphs that frame and provide orientation to a story, in the journalistic voice's explanations and contextualisation, in quotations from experts or in arguments in debate articles. This variety of uses makes scientification rather free-floating in terms of its content. As visible in the examples provided, scientification can be used to highlight certainty and thereby provide weight to an argument, or, in contrast, it can be used to highlight scientific uncertainty. Furthermore, there are different "sources" for scientification, which have different knowledge, understanding and experience of specific issues.

This insight takes us to another insight that the case of artificial turf provides, namely that scientification is an important weapon in the struggle over environmental risks. Central in CDA is the understanding of the cultural sphere, which the media is part of, as locus of debate and contestation between different power interests and ideologies (Fairclough, 1995). The case of artificial turf, and the role of science in it, is no exception. For the tyre industry, the continuous use of gum granules from worn out tyres is important, and being able to point to scientific uncertainty in relation to the environmental harms of artificial turf pitches is therefore equally important. For environmental activists, politicians or local government officials who are guided by precaution, any statistics pointing to the environmental dangers of artificial turf pitches are worrisome and will probably be used as evidence for the certainty of the risk. For researchers, who have the difficult task of being fact bearers, it may be important to stress what is known, but also what is not known, and what needs to be done to know more. All of these actors, the interests they represent, and the institutional logics of which they are part shape the use of scientification in relation to this specific environmental risk. On the one hand, this diversity shows that science is an important realm for attaching weight to an argument, or to explain something, despite societal tendencies towards post-truth and science neglect (McIntyre, 2018). On the other hand, this same diversity points to the possible relativisation and watering down of science in media discourse.

That insight takes us to a third and final insight, which is the risks of scientification in media discourse. Due to the use of science for these different purposes and to uphold different interests, the discursive use of science in the media must be guarded. Otherwise, we may face a situation where anything goes, and allusions to science can be used to uphold doubtful claims or even to spread misinformation, which would help strengthen current post-truth tendencies.

In the task of guarding science, newsrooms have a big responsibility. At the end of the day, it is the newsrooms that decide what to publish, not the researchers or the actors being cited, or the universities or other organisations who reach out to the media for attention. The examples provided in this chapter suggest that there is room for improvement in the use of science by the media. Journalists and editors could, in some cases, be more critical of the use of scientification by some social actors and also by themselves. This would perhaps require a deeper understanding of science in newsrooms, something that is not encouraged in a situation where science journalism is under pressure (Schäfer, 2017). However, a key problem, which is not new, lies in the recontextualisation of science and environmental issues, and the adaption of science and the environment to the logic of contemporary news media (see Berglez, 2011; Olausson & Berglez, 2014). Science needs to fit the formats of the news media, and in a commercialised landscape, it needs to fit stories that are sellable. As Karidi (2018) points out, commercial media logics serve stories focused on, among other things, conflict and negativity, which if translated to the coverage of environmental issues, would foster a focus on crises and problems. Here, figures such as the ones provided by IVL on microplastics, regardless of the uncertainty around them, give the media the foundations for constructing an environmental problem. Being commissioned by the SEPA, and being part of governmental environment policies, also attributed weight to IVL's figures. Researchers pointing to uncertainty and the scarce state of sure facts, or highlighting results of specific case studies that perhaps are difficult to generalise, are then discursively made to fit an already defined problem-frame. When interviewing stakeholders about communicating the environmental risks of microplastics and artificial turf, we found that experts point to difficulties in addressing scientific uncertainty with journalists (Abalo & Olausson, 2023). There is fear about addressing uncertainty with journalists because statements could be used to amplify risks and dangers. Some of the examples presented in this chapter make such scepticism towards the media understandable.

The insights in this chapter suggest that a change to the discursive treatment of science in the media, when dealing with environmental risks, would require changes not only in newsroom practices but also in the commercial logics that guide much of the news media, especially in the Western world. Although large and profound, such changes are relevant to consider in order to strengthen the credibility of the news media's treatment of science in the context of environmental risks.

Notes

1 All excerpts have been translated from Swedish by the author.

2 SDAB commissioned three case studies by the School of Education and Communication, Jönköping University, on Swedish news media coverage on artificial turf as an environmental problem. PI and academic responsible for that project was Professor Ulrika

Olausson. The studies were designed by the author and the PI, in cooperation with the company. An agreement between the parties was written which declares that the commission and funding do not impact the openness, independence and objectivity of the study and which allows the researchers to publish results without the consent of the funder. Therefore, there are no conflicts of interest. This chapter was written without funding from SDAB or any other organisation.

3 The Swedish terms that are used for research (*forskning*) and researchers (*forskare*) mean scientific research. These are not to be confused with the use of the Swedish word *research* (borrowed from English), which signifies journalistic research.

References

Abalo, E. (2019a). *"Fotbollsplanen kan vara en tickande miljöbomb": om konstruktionen av risk och expertis i svenska nyhetsmediers rapportering om konstgräs som miljöfråga.* Research Reports No. 014. Jönköping University, School of Education and Communication. Jönköping, Sweden. http://urn.kb.se/resolve?urn=urn:nbn:se:hj:diva-47070

Abalo, E. (2019b). *Konstgräs som miljöfråga i svenska nyhetsmedier: en kvantitativ innehållsanalys.* Research Reports No. 10. Jönköping University, School of Education and Communication. Jönköping, Sweden. http://urn.kb.se/resolve?urn=urn:nbn:se:hj:diva-43582

Abalo, E. (2021). Between facts and ambiguity: discourses on medical cannabis in Swedish newspapers. *Nordic Studies on Alcohol and Drugs, 38*(4), 345–360. doi:10.1177/1455072521996997

Abalo, E. & Olausson, U. (2023). An environmental problem in the making: how media logic molds scientific uncertainty in the production of news about artificial turf in Sweden. *Journal of Science Communication, 22*(1), 1–18. https://doi.org/10.22323/2.22010201

Allan, S., Anderson, A., & Petersen, A. (2010). Framing risk: nanotechnologies in the news. *Journal of Risk Research, 13*(1), 29–44. https://doi.org/10.1080/13669870903135847

Altheide, D.L. (1997). The news media, the problem frame, and the production of fear. *The Sociological Quarterly, 38*(4), 647–668. doi:10.1111/j.1533-8525.1997.tb00758.x

Altheide, D.L. & Snow, R.P. (1979). *Media logic.* Beverly Hills: Sage.

Backhaus, T., Wagner, M. (2020), Microplastics in the environment: much ado about nothing? A debate. *Global Challenges, 4*, 1900022. https://doi.org/10.1002/gch2.201900022

BEKOGR [Kommunernas konstgräsnätverk] (2023). Välkommen till Kommunernas konstgräsnätverk. https://bekogr.se/

Berglez, P. (2011). Inside, outside, and beyond media logic: journalistic creativity in climate reporting. *Media, Culture & Society, 33*(3), 449–465. doi:10.1177/0163443710394903

Boykoff, M.T. & Boykoff, J.M. (2004). Balance as bias: global warming and the US prestige press. *Global Environmental Change, 14*(2), 125–136. doi:10.1016/j.gloenvcha.2003.10.001

Burton, G.A. (2017). Stressor exposures determine risk: so, why do fellow scientists continue to focus on superficial microplastics risk? *Environmental Science & Technology, 23*(51), 13515–13516. doi: 10.1021/acs.est.7b05463

Burton, G.A. & Cervi, E.C. (2019). Environmental stressor importance: science versus media. *Environmental Toxicology and Chemistry, 38*(12), 2587–2592. https://doi.org/10.1002/etc.4606

Carvalho, A. (2005). Representing the politics of the greenhouse effect: discursive strategies in the British media. *Critical Discourse Studies, 2*(1), 1–29. https://doi.org/10.1080/17405900500052143

Carvalho, A. (2008). Media(ted) discourse and society. *Journalism Studies, 9*(2), 161–177. https://doi.org/10.1080/14616700701848162

Dovbysh, O., & Perkiömäki, M. (2025). Journalists-sources relations in Russian environmental journalism. In M. M. Roslyng, A. Rantasila, & A. M. Jönsson (Eds.), *Communicating science, climate change and the environment in hybrid media: Constructed facts, contested truths* (pp. 216–232). New York: Routledge

ECHA. (2019). *Annex XV restriction report: proposal for a restriction.* European Chemicals Agency. https://echa.europa.eu/documents/10162/05bd96e3-b969-0a7c-c6d0-441182893720

ECHA (n.d.). Microplastics. ECHA. Retrieved April 24, 2023, from https://echa.europa.eu/hot-topics/microplastics

Fairclough, N. (1995). *Media discourse.* London: Hodder Education.

Fairclough, N., & Wodak, R. (1997). Critical discourse analysis. In T. van Dijk (Ed.), *Discourse studies: a multidisciplinary introduction* (Vol. 2, pp. 258–284). London: Sage.

Farkas, J., & Schou, J. (2020). *Post-truth, fake news and democracy: mapping the politics of falsehood.* New York: Routledge. https://doi.org/10.4324/9780429317347

Höijer, B. (2009). Risk. In P. Berglez & U. Olausson (Eds.), *Mediesamhället: centrala begrepp* (p. 241–264). Lund: Studentlitteratur.

IPCC (2023). *Synthesis report of the IPCC sixth assessment report (AR6).* Intergovernmental Panel on Climate Change. https://report.ipcc.ch/ar6syr/pdf/IPCC_AR6_SYR_LongerReport.pdf

Jönsson, A. M. (2025). Mediated science and issues of public knowledge and trust. In M. M. Roslyng, A. Rantasila, & A. M. Jönsson (Eds.), *Communicating science, climate change and the environment in hybrid media: Constructed facts, contested truths* (pp. 153–175). New York: Routledge

Jung, D. (2015). Sociology, protestant theology, and the concept of modern religion: William Robertson Smith and the "scientification" of religion. *Journal of Religion in Europe, 8*(3–4), 335–364. https://doi.org/10.1163/18748929-00804006

Karidi, M. (2018). News media logic on the move? In search of commercial media logic in German news. *Journalism Studies, 19*(9), 1237–1256. doi:10.1080/1461670x.2016.1266281

Krång, A.-S., Olshammar, M., Edlund, D., Hållén, J., Stenfors, E. &Winberg von Friesen, L. (2019). *Sammanställning av kunskap och åtgärdsförslag för att minska spridning av mikroplast från konstgräsplaner och andra utomhusanläggningar för idrott och lek.* IVL Svenska Miljöinstitutet. Stockholm, Sweden. www.diva-portal.org/smash/get/diva2:1549597/FULLTEXT01.pdf

Lambert, S., & Wagner, M. (2018). Microplastics are contaminants of emerging concern in freshwater environments: an overview. In M. Wagner & S. Lambert (Eds.), *Freshwater microplastics: emerging environmental contaminants?* (s. 1–24). Cham: Springer.

Leidecker-Sandmann, A.P., Schütz, A., & Lehmkuhl, M. (2022). Selected by expertise? Scientific experts in German news coverage of COVID-19 compared to other pandemics. *Public Understanding of Science (Bristol, England), 31*(7), 847–866. https://doi.org/10.1177/09636625221095740

Lidskog, R. & Olausson, U. (2013). To spray or not to spray: the discursive construction of contested environmental issues in the news media. *Discourse, Context & Media, 2*(3), 123–130. doi:10.1016/j.dcm.2013.06.001

Magnusson, K., Eliasson, K., Fråne, A., Haikonen, K., Hultén, J., Olshammar, M., Voisin, A. (2016). *Swedish sources and pathways for microplastics to the marine environment: a review of existing data.* Number C183. IVL Swedish Environmental Research Institute. Stockholm, Sweden.

Magnusson, K., Eliasson, K., Fråne, A., Haikonen, K., Hultén, J., Olshammar, M., Voisin, A. (2017). *Swedish sources and pathways for microplastics to the marine environment: a review of existing data.* Updated version. IVL Swedish Environmental Research Institute. Stockholm, Sweden. www.ivl.se/english/ivl/publications/publications/swedishsources-and-pathways-for-microplastics-to-the-marine-environment.html

Males, J. & Van Aelst, P. (2021) Did the blue planet set the agenda for plastic pollution? an explorative study on the influence of a documentary on the public, media and political agendas, *Environmental Communication, 15*(1), 40–54, DOI: 10.1080/17524032.2020.1780458

McIntyre, L.C. (2018). *Post-truth.* MIT Press

Musi, E., & Reed, C. (2022). From fallacies to semi-fake news: improving the identification of misinformation triggers across digital media. *Discourse & Society, 33*(3), 349–370. https://doi org.proxy.library.ju.se/10.1177/09579265221076609

NBC News. (2019, September 18). Greta Thunberg To Congress: 'I Want You To Listen To The Scientists' [Video]. YouTube. www.youtube.com/watch?v=vznT5nVOZgw

Olausson, U. (2009). Global warming—global responsibility? media frames of collective action and scientific certainty. *Public Understanding of Science, 18*(4), 421–436. https://doi-org.proxy.library.ju.se/10.1177/0963662507081242

Olausson, U. & Berglez, P. (2014). Media and climate change: four long-standing research challenges revisited. *Environmental Communication, 8*(2), 249–265. doi:10.1080/17524032.2014.906483

Peters, H.P., & Dunwoody, S. (2016). Scientific uncertainty in media content: introduction to this special issue. *Public Understanding of Science, 25*(8), 893–908. https://doi-org.proxy.library.ju.se/10.1177/0963662516670765

Rantasila, A. (2025). Constructing trust with affective discipline: Finnish nuclear energy experts and the Fukushima Daiichi disaster. In M. M. Roslyng, A. Rantasila, & A. M. Jönsson (Eds.), *Communicating science, climate change and the environment in hybrid media: Constructed facts, contested truths* (pp. 176–193). New York: Routledge

SAPEA. (2019). *A scientific perspective on microplastics in nature and society.* Science advice for policy by European academies. https://doi.org/10.26356/microplastics

Schäfer, M.S. (2009). From public understanding to public engagement: an empirical assessment of changes in science coverage. *Science Communication, 30*(4), 475–505. https://doi-org.proxy.library.ju.se/10.1177/1075547008326943

Schäfer, M.S. (2017). How changing media structures are affecting science news coverage. In K.H. Jamieson, D.M. Kahan, D.A. Scheufele (Eds.), *The Oxford handbook of the science of science communication (Electronic ed.).* Oxford University Press. 10.1093/oxfordhb/9780190497620.013.5

Slovic, P. (1999). Trust, emotion, sex, politics, and science: surveying the risk-assessment battlefield. *Risk Analysis, 19*(4), 689–701. https://doi.org/10.1111/j.1539-6924.1999.tb00439.x

Völker, C., Kramm, J., & Wagner, M. (2020). On the creation of risk: framing of microplastics risks in science and media. *Global Challenges, 4*(6), 1900010–n/a. https://doi.org/10.1002/gch2.201900010

von Stuckrad, K. (2014). *The scientification of religion: an historical study of discursive change, 1800–2000*. De Gruyter. https://doi.org/10.1515/9781614513490

Yamaguchi, T. (2016). Scientification and social control: defining radiation contamination in food and farms. *Science, Technology & Society, 21*(1), 66–87. https://doi.org/10.1177/0971721815622741

3 Web of denial

Climate change denial discourse on Instagram

Virág Vécsey

Introduction

Although a strong scientific consensus exists on anthropogenic climate change (Cook et al., 2016), a not insignificant minority denies it. Previous research has examined the issue concerning various demographic indicators and found that political-ideological attitudes correlate most clearly with climate beliefs (Poortinga et al., 2019). Right-wing populist political forces in both the US and Europe are obstructing the mitigation of the consequences of climate change, as their leaders and followers are often climate sceptics (Gomes & Böhm, 2022; Kulin, Seva, & Böhm, 2021; Lockwood, 2018). Fundamentally anti-elitist and anti-establishment populist rhetoric, nationalist ideology against global-scale political action, and biassed critiques of capitalism are the main reasons for this. Clearly, the problems with climate change communication are related to a structural crisis in liberal democracies (Forchtner, 2019).

A shift towards social media can be observed concerning climate change and science communication in a broader sense (Treen, Williams, & O'Neill, 2020). Certain features of social media, such as its openness and speed of information dissemination through user sharing, accelerate and facilitate the spread of misinformation and disinformation (Wu et al., 2016). In the case of Instagram, such features include fewer credibility cues due to the focus on visual content and rapid information processing with less time for checking veracity (Vraga et al., 2020). In social media, disinformation is spread by algorithmic bias, malicious bots, spammers, individual users, organisations or astroturfers. The latter term refers to large groups of paid users hired to disseminate malicious information on social media (Treen, Williams, & O'Neill, 2020). Research on climate change denial's appearance on social media has shown how scientific information is manipulated and falsified to adopt the appearance of credibility (Anderson, 2017; Björnberg et al., 2017; Dunlap et al., 2016). Climate change is among the most important topics impacted by misinformation, which significantly affects public attitudes towards the issue (van der Linden et al., 2017).

This chapter aims to provide insights into the climate denial discourse on social media, with a particular focus on its visual aspects. Thus using methods of multimodal critical discourse analysis and content analysis, this research examines

DOI: 10.4324/9781003479550-4

the top 300 posts on Instagram tagged #climatehoax or #climatechangehoax. Combining qualitative and quantitative methodologies enables us to examine the communicative act as a whole, paying attention to textual and visual content, the ideological framework, and the underlying political and power structures in which climate denial is embedded. The main question of this study is: What ideological framework is climate denial discourse embedded in and how is the topic represented visually in social media? The way people think about climate change and their perceived role of responsibility determines public policies in relation to climate change mitigation and adaptation. In the age of social media this is increasingly being communicated visually. The web of diverse conspiracy theories linked to climate denial identified in this research is being effectively exploited by the far right to obstruct environmental policy making on climate change mitigation and adaptation, one of the biggest and most pressing issues of our time. The findings of this research will contribute to understanding the patterns, ideologies and power structures behind the communication of climate denial, and could help to develop strategies to counter them.

Interestingly, even though every content analysed was tagged #climatehoax or #climatechangehoax, less than half of the posts related to climate change and its denial, and only 22% did so visually. This is because climate change denial is presented in social media as an element of a well-defined set of conspiracy theories that are politically and ideologically associated with the far right. These conspiracy theories can be linked to the denial of basic scientific facts, the questioning of political elites and organisations of military, economic or social power (e.g. NATO, UN), the rejection of liberal values and hatred of certain religious minorities. Based on my findings, I argue that even if climate denial is formulated pictorially on Instagram, it is mostly through the reuse and recycling of well-known imagery of climate change, such as memetic versions of Greta Thunberg, pictures of weather events and images accompanying articles from news media. Thus visual communication is not creating and using an original iconography, and contains climate denialism only in captions or other added textual elements.

Before introducing the case, selection process and research methods used, I will briefly address previous research findings on climate scepticism, its relation to political forces and ideologies, and its communication through social media.

Climate scepticism and the (far)right

Climate sceptics are not a homogenous group. Various studies group climate sceptics differently according to their attitudes to the scientific arguments, the political and economic implications of addressing the issue, and the severity of climate change. According to Poortinga and colleagues (2019), scepticism can be related to climate change's trend, attribution or impact. Björnberg (2017) adds to this the fourth category of consensus, which labels the radical group of climate deniers, who simply question the existence of climate change. In this corpus, the latter predominates, arguing that climate change has no anthropogenic causes or does not exist but is merely the fabrication of alarmist environmentalists and an elusive political elite.

Research has shown that conservative activists are less engaged with the topic of climate change and much more likely to share climate scepticism than liberals (Carvalho, 2007; Dotson et al., 2012; McCright, Dunlap, & Marquart-Pyatt, 2016; Painter & Gavin, 2016; Barkemeyer et al., 2017; Han et al. 2017). Examining the climate sceptic communication of conservative think tanks, Boussalis and Coan (2016) found that it has grown exponentially in the 15 years studied and goes beyond questioning the scientific consensus to challenging the legitimacy of policies. When exploring the roots of right-wing populist forces' climate scepticism, the structuralist approach and the ideological explanation are considered (Lockwood, 2017). Research examining the significance of nationalism in the political ideology of the right concludes that a doctrine that emphasises national sovereignty is fundamentally opposed to taking transnational action against climate change as a global issue (Huber, 2020; Kulin, Johansson, & Dunlap, 2021). In Western countries, research has found that people's political affiliations primarily determine public perceptions of climate change. In Anglophone countries, the Democrat-Republican, or right-left divide on climate change is more potent than in Europe, where a milder polarisation of this type mainly pertains to Western Europe (Kulin, Johansson, & Dunlap, 2021). In the US in particular, but also in other Anglophone countries, climate scepticism is the flagship of science scepticism and science denialism, which is promoted by politically motivated groups with powerful financial support (McCright & Dunlap, 2011, Van der Linden et al., 2017). This network of actors, also called the denial machine (Elsasser & Dunlap, 2013) and the Merchants of Doubt (Van der Linden et al., 2017), is divided by Björnberg (2017), into six partially overlapping categories, namely scientists, government, media, industry, political and religious organisations and the public. The anthropogenic causes or mere existence of climate change is denied by a small group of scientists, who are typically not part of the scientific community, often not belonging to scientific associations but working for think tanks. Unsurprisingly, science questioning anthropogenic climate change is propagated by industrial actors interested in coal and oil mining and car manufacturing (Björnberg, 2017). This campaign-like tide of misinformation is also prominent in online social media, contributing to the depoliticisation and polarisation of climate change opinion and hindering the increasingly urgent action to mitigate its consequences.

Previous research on social media communication of climate change has disproportionately focused on platforms centring around textual information, particularly on Twitter (Pearce et al. 2019). These studies examined communication sources and the phenomena of polarisation, echo chambers and climate scepticism. They concluded that although there has long been concern that social media is reshaping the media structure in such a way that the role of professional news sources is diminishing, mainstream news sites remain the most dominant sources in climate change communication when considering external sources cited on Twitter (Newman, 2017). When considering the sources of re-shared posts within Twitter, the picture changes and laic users (i.e., users unrelated to science or media) are in the majority. Research by Williams and colleagues (Williams et al., 2015) used network analysis to examine the spread of five climate change-related hashtags on Twitter. It concluded that the most active Twitter users tend to be both climate activists and sceptics, with a tendency to form solid opinions and cluster in like-minded groups, leading to the polarisation of communication. By analogy, the corpus analysed presumably belongs to a vociferous minority on Instagram.

Cases, Data, and Methods

Case selection & data

My research aims to explore the communicative acts used by climate sceptics on Instagram. Visual content consumption, production and sharing is common on social media. Despite this, there are but a few examples (Allgeier, 2019 in the case of YouTube; Hermann, Rhein, & Dorsch, 2022; Vraga et al., 2020 in the case of Instagram) when researchers investigated YouTube and Instagram content in the field of climate change communication (Björnberg et al., 2017; Pearce et al., 2019; Treen, Williams, & O'Neill, 2020) and none have focused solely on climate scepticism or denial. Gibbs and colleagues found that while there are many similarities between platforms in social media (e.g. use of hashtags, sharing and following features, ability to create user accounts, etc.), there is a specific grammar, style and logic unique to the platform. The notion of platform vernacular extends the characteristic features of social media, such as self-representation and culture of participation described by Burgess and colleagues (2006), to the possibilities of creative expression. The platform binds creative expression; thus, vernacular can be considered an extension of focus from the platform's policies and politics to the user language. The vernacular language of the platform is defined by the software and hardware that ensure the functioning and use of the social media platform, but the communication habits and practices of the users also shape it. Initially, the platform was associated with thematic and aesthetic determinants such as the dominance of selfies and influencers and the staging aesthetic and use of photo filters (Rogers, 2021). However, we will see that neither of these features of the platform vernacular is present in the subculture of Instagram studied here. From a technical point of view, Instagram differs from other social media platforms in that it is designed specifically for mobile use, both in the case of the recipient and the content producer, which, together with the dominance of images, implies a fast, relatively superficial reception. Content for Instagram is designed for fast scrolling due to the vertical orientation of the platform's design. These characteristics can influence the way users perceive information and misinformation. The factor of credibility plays a different role in superficial, fast and primarily visual communication than in verbal information. The former does not leave room for intellectual reflection, but rather communicates on an emotional basis, which in our case is effectively achieved through humour and satire.

As a visual social platform that has become extremely popular since its launch in 2010, Instagram had 1.21 billion active users in 2021 (Dixon, 2023). Its popularity provides reason enough to consider the content shared here when considering the spread of misinformation. Research on political communication has shown that during the 2016 US presidential election, this platform and other social media sites were involved in the Russian disinformation campaign. This makes it clear that content spread here goes well beyond aestheticising images, for which the platform initially was created (Howard et al., 2019) and has become a visual political marketing tool (Highfield & Leaver, 2016). In addition to influencers, brands and individual users, Instagram now features media organisations with institutional accounts. When designing this research, I have chosen not to filter users to include views and communication practices of far-right individuals who are sceptical towards larger right-wing media organisations (e.g. Fox News, Breitbart, Prager u) and these professional content creators. Thus, I filtered hashtags for explicitly climate sceptical content (e.g. *#climatescam, #climatehoax, #climateagenda, #climatechangescam, #climatefraud* etc.) and looked at the two most commonly used hashtags: *#climatechangehoax* and *#climatehoax*[1].

These tags led to content mainly uploaded by individual users and thus provided diversity in the production site of communication. However, the network analysis of content producers was not part of this research. Data was collected manually on 22.04.2023 and 24.04.2023. I analysed the first 300 posts of the *top* results of each hashtag (n = 600). Instagram offers the "top" hashtags alongside the most "*recent*" content and "*reels*". Since 2016, Instagram, like many other social media platforms, has shifted from displaying posts chronologically to displaying them in an order sorted by the algorithm. This makes the underlying logic to rankings, such as the *top* feed, unintelligible, but this option was the most appropriate for the research as a default setting, which users most likely encounter. Algorithmic rankings like this are not without human input, and they are formed according to the intentions of many actors. Ranking culture (Rieder et al., 2017), characteristic of social media, reinforces popularity and, in this sense, is descriptive, performative and affirmative.

Research methods

To gain a deeper understanding of climate sceptic communication, this research combined quantitative content analysis with the qualitative methods of multimodal critical discourse analysis. This research method enables us to examine the communicative acts in their complexity as it seeks to explore how discourse sustains, counters or rebuilds social and political power relations within which it appears (Fairclough, 2015 [1989]). As we shall see, this proves particularly useful for investigating the ideological embeddedness of this community, tying together climate denialism with conspiracy theories circulated by populist right-wing political forces. When dealing with the highly politicised topic of climate change, it is essential to investigate underlying power structures shaping the discourse around it. Hence despite being a global-scale problem, climate change unevenly impacts society. The multimodal approach was necessary to examine the content, including visual, textual and aural elements. Even though Instagram centres on visual communication, images are always accompanied by tags and often captions. The latter depends on the time, interests and preferences of the users, but it can be said that the verbal context of the shared image cannot be ignored, as well as "aural" features such as the configuration, functions and design of the platform, which I have already addressed concerning the platform vernacular. This multimodal ensemble collectively shapes meanings (Domingo, Jewitt, & Kress, 2020).

First content analysis was used to code the posts into the most frequently occurring categories according to their formal and semantic features and the source of the content. Differentiating content along formal features meant distinguishing posts with no pictorial elements (i.e. text only (12%), memes (17%), other posts with images (65%) and the variety of videos (6%)). The latter was not included in the analysis any further as this first step of coding as it would have required a different research method.

Communicative sources-built the following categories: re-shared news (10%), social media materials from another platform (7%) and content that has not been added from another source (i.e. original content: 82%[2]). Considering communicative sources was essential to understand how main actors of power structures, such as the media industry, affect structures in discourse. Accounting for these is crucial from the perspective of critical discourse analysis (van Dijk, 2002).

It would not have been possible to categorise the posts based solely on their iconography, as many of them could only be interpreted in conjunction with their caption or textual elements on the image. In terms of the results, this means that the platform specialised in pictorial storytelling is frequently being used to share text messages in this case. Thus semantic features of posts built a set of categories (see Figure 3.1.) such as pictures of the sky with kerosene strips (i.e. chemtrail category),

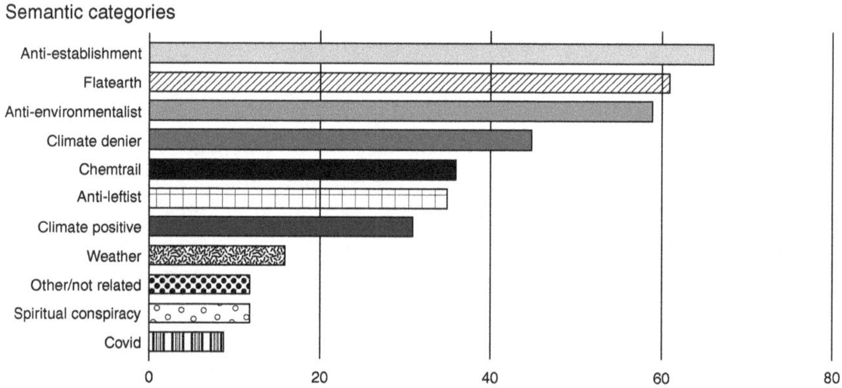

Figure 3.1 Semantic categories of the top 300 posts tagged either #climatechangehoax or #climatehoax (excluding videos and content created by the "NFT artist").

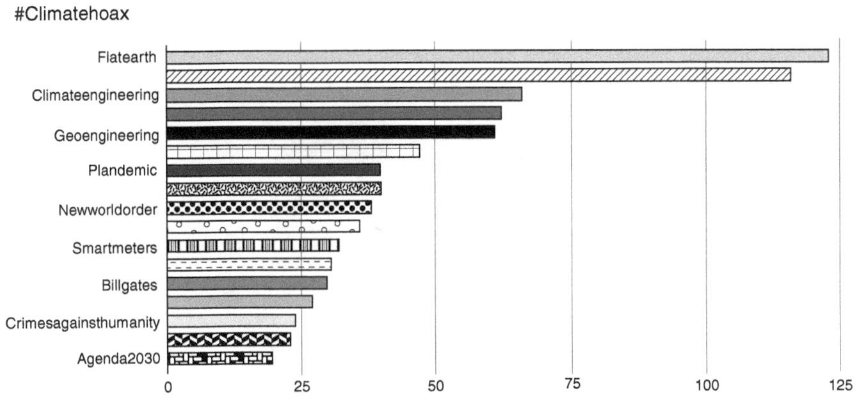

Figure 3.2 Most often used categories of hashtags accompanying the top 300 post results tagged #climatehoax.

posts about a flattened globe or an artificial, "fake" space (i.e. flat-earth category). Posts that, in line with populist narratives, draw attention to the machinations of a money-hungry elite that is attacking the lives of ordinary citizens, changing the weather, feeding the people bugs etc. were tagged anti-establishment, posts which were explicitly against socialism or communism or democrats (i.e. the political party in the US) were tagged anti-leftist and content humiliating, ridiculing or attacking environmental activists and their issues (e.g. use of renewable energy, electric vehicles) were tagged anti- environmentalist. Content claiming climate change doesn't exist because of certain weather events (e.g. it's cold or snowy) can be considered a subcategory of posts labelled climate denier. The latter label was used for content which centred on climate denial as opposed to posts which barely touched on it on the level of hashtags.

Multiple images added to a single post were treated as one image in case of homogeneity, otherwise as many as the number of categories they belonged to. For posts that could be classified in multiple categories simultaneously the most characteristic category was chosen.

Second, the content analysis of hashtags associated with the posts tagged *#climatehoax* was based on an analogous principle, where hashtags with similar meanings were grouped into a single code category to make the results more readable. For example, the tags *#stopspraying, #fuckchemtrails, #toxicsky, and #chemsky* all fell into the chemtrail category. I set up a total of 70 categories, of which Figure 3.2. represents the 14 categories with at least 20 hashtag occurrences. Here it is important to note that some uploaders add long lists of hashtags to all their posts, regardless of their visual content, to increase their posts' views. Thus, a purely quantitative analysis of the tags shows distorted results, all the more so because hashtags are used in a way that is both ironic and informative, with many posts being tagged *#climatehoax* and *#climatechange* at the same time or, by

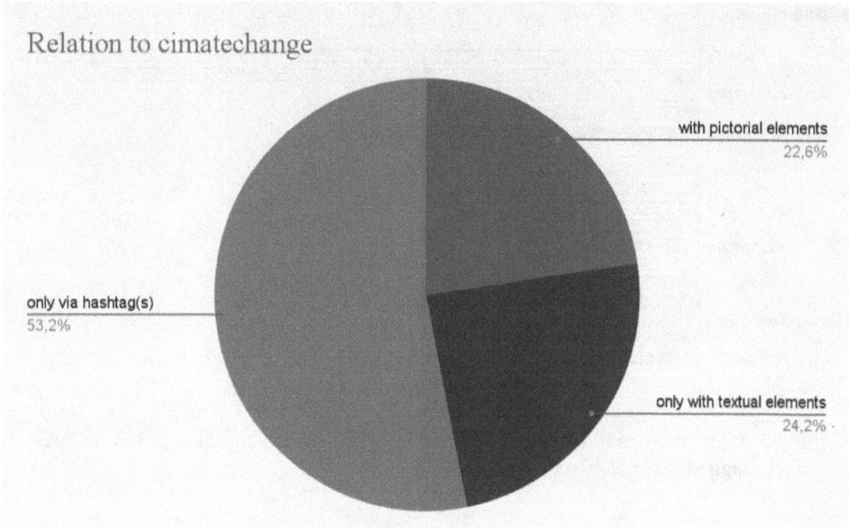

Figure 3.3 Posts relation to climate change (videos and climate-positive content excluded).

analogy, *#fossilfuels:D* and *#greenenergy* etc. Climate-positive[3] posts, dominated by vegan activist content and accounted for 6% of the sample analysed, were not included in the further analysis.

After coding and carefully reviewing the sample, it was striking how many posts do not address climate change beyond one or two hashtags, either visually or textually. Thus, as a third step, I categorised each post according to whether it was related to climate change denial (scepticism) beyond the two hashtags analysed and whether it was purely textual or visual (see Figure 3.3.). This analysis didn't include videos and climate-positive content in line with the above. To get precise results here, I have grouped each post into a single category, typically the most relevant in terms of content.

Posts that were visually related to climate denial constituted less than a quarter of the sample (see Figure 3.3.)

In addition to reconstructing the meaning of mediatised representations, critical discourse analysis examines the social, political and power relations embedded in their production, dissemination and reception (Fairclough, 2015; Stahl & Salam, 2015). Accordingly, linking the semantic, ideological and communicative elements helped me explore the underlying power structures of the climate denial discourse and investigate questions such as: Is the uploader's country of origin deductible? Is there any indication of political ideology or even party-political preferences? If content from a news media platform is being re-shared, to which political side is it attributable? Which economic, political, and ideological forces are the public figures and politicians in the image associated with? The starting point of critical discourse analysis is that both visual and verbal discourses are social constructs that shape society and at the same time are being shaped by society (Machin & Mayr, 2012). Behind seemingly neutral language and taken-for-granted assumptions, power interests can be revealed. In our case, according to the populist rhetoric of conspiracy theories, such self-evident forms are the juxtaposition of the powerful "them" and the well-aware "us". Nowhere in this paranoid theory of global intrigue, however, is the place in the power structure of those who share the theory, or even the media platform they use, called into question.

Analysis

Considering hashtags and visual semantic features (see Figures 3.1. and 3.2.), it is clear that the empirical sample is embedded in a broader set of conspiracy theories alongside climate change denial. In social media, hashtags mark the content of posts and act as searchable and connective elements to embed the given context in a network of ideologically, content-wise or network-wise related posts and posters. All 14 hashtag categories (e.g. chemtrails, flat Earth, depopulation, globalist agenda etc.) used at least by 20 posts are linked to far-right conspiracy theories. In addition, as Figure 3.3. shows, less than half of the empirical sample refers to climate denial more directly than through the use of a few hashtags, which is a sign that climate denial is often just one in a series of conspiracy theories that users attach to their posts as a matter of course. In line with these findings, the first part of the analysis will focus on the web of conspiracy theories drawn by the corpus, and the second will examine the minority of the corpus, which visually represents climate denial.

Negative referentiality- the web of conspiracy theories

Conspiracy theories are attempts to explain significant social and political events, factors and circumstances by the covert plotting of some ill-intentioned group. The conspirators can be lead politicians and any other group with perceived or actual power (Douglas et al., 2019). Besides the government, Bill Gates' name appears most frequently in the sample studied, but also organisations of global importance such as the UN, its sub-organisation, the WHO (World Health Organisation) and the WEF (World Economic Forum). It is worth noting that the latter three tags appeared mostly together, forming one category. The malevolent "they" of conspiracy theories hardly seems present in the visual language of the posts, except for Greta Thunberg and Klaus Schwab (WEF President), whose face occurs in a small fraction of the sample. Thus, the binary opposite categories of the powerful elite, "*them*" versus "*us*", which define populist rhetoric, are illuminated by references in the form of hashtags. Conspiracy theories relate to populist politics and ideas (Butter, 2022). In addition to organisations of the UN, conspiracy theories embedded in the sample examined also named Bill Gates (see Figure 3.4.), liberals, democrats, big pharma and better-known Jewish stakeholders such as Rockefeller and George Soros in line with well-known antisemitic, far-right populist narratives. Global in nature, these theories have a national dimension only in the case of

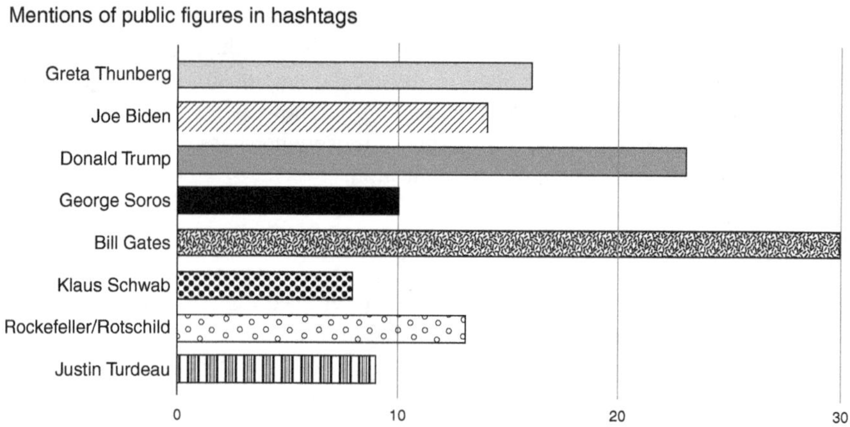

Figure 3.4 Mentions of public figures in hashtags.

the US and Canada. Among politicians, Donald Trump is the most frequently mentioned, usually referred to as a positive character (e.g. *#bringbackTrump, #DonaldTrumpisourPresident, #Trump2024*). Joe Biden and Justin Trudeau are also frequently mentioned but with negative connotations (e.g. *#Trudeaumustgo #traderjoe #bidenflation #notmypresident*).

Also linked to the US is the Democrat-Republican dichotomy, of which the former appears, always demonised (e.g. *#democratshateamerica, #decmocratsdestroyamerica, #democraticcoruption*), whereas the *#makeamericagreatagain* type of tags is linked to Donald Trump. It is important to note that the word Republican does not appear in the tags, but conservative does. The focus on the US of these hashtags can be due to the worldwide use of the English language, but the international nature of the chemtrail posts with a large number of Slovenian and British uploads suggests that mainstream international interest in US domestic politics may be driving the corpus. Therefore if the evil other has a nationality, it must be American, but most often, it is an undefined actor of the global political and economic elite. It can also be argued that ideologically, the uploaders identify themselves as conservative, opposing liberalism, globalism and environmentalism.

Consequently, minority identity politics is also being attacked (e.g. #homosexualagenda, #wokeagenda) and occurs in the web of conspiracy theories defined by tags among posts whose subject is unrelated to this. Sometimes socialism and communism appear in the corpus, either as tags or a few times in the form of images, always with negative connotations. In one post, even the symbol of the sickle and the hammer is depicted; in another, Marx occurs as a meme, which pictures the idea of climate hysteria as the tip of the iceberg of Marxist ideology. However, these historical and philosophical references are rare as outsiders to the mainstream of popular culture. The general distrust of experts and politicians is a peculiarity of conspiracy theories, and far-right ideological groups often propagate conspiracy theories to promote their views (Fuchs, 2021; Ekman, 2022; Butter & Knight, 2020).

Recently, conspiracy theories have drifted from the media periphery into the mainstream (Butter & Knight, 2020). Reasons for this include the transformation of the media landscape, i.e. the rise of social media and the weakening of the traditional gatekeeper role. In the online space, social media platforms legitimise conspiracy theories discursively and interactively through likes and shares and technically through algorithmic processes (Ekman, 2022). However, as Fuchs (2021) points out, social media is not a generator of conspiracy theories, merely a carrier of symptoms of a highly exposed society prone to these. This, he argues, stems from the negative dialectic of capitalist society. Content uploaders in the sample examined are not organisations but individual account users. However, due to a lack of resources, this research did not include a network analysis of the users, so it cannot give an idea of whether there are any and, if so, which interest groups are behind the distribution of misinformation and disinformation. The corpus as a whole suggests that anthropogenic climate change deniers do not only deny science in this respect but also reject the existence of gravity (*#gravityisfake*), the sphericity of the Earth (10% of visuals, 15% of hashtags, i.e., *#flatearth # itsflat,* etc.) and do not believe in the existence of space (*#spaceisfake*). Contradictory denying scientific facts is often coupled with seemingly rational counterarguments, a phenomenon Fuchs calls irrational rationality (Fuchs, 2021, pp. 72), a marker of conspiracy theories. This is also evident in the 2% of posts that use some data visualisation for the sake of credibility. An example of this is a graph showing the composition of air, which is intended to show that since there is much more nitrogen and oxygen in the air than carbon dioxide, there is no carbon emission-induced climate change.

Mostly, memes popularise the flat-earth theory and the 'chemtrail' conspiracy. Flat Earth memes recycle the globe, one of the most commonly used visual elements in climate change communication and environmental communication in general, albeit with sceptic captions or in a flattened form. Under the chemtrail conspiracy (6% of visuals, 14% of hashtags), a primarily unspecified "*they*" inject aerosols containing various metals from planes into the air that obscure the sun, are toxic and cause vitamin D deficiency. The theory is illustrated in the sample examined with images of slightly cloudy skies or skies with a kerosene streak from an aircraft, but it also takes a surprisingly diverse form in the meme version (see Figure 3.5.). Even though the sample is politically leaning to the (far) right, there is ambivalence about the political nature of chemtrail conspiracy. While fears of a new world order and overly powerful governments are indeed associated with the right, environmental concerns and objections against corporations are more tied to the political left (Cairns, 2016).

It is not only the accompanying texts appearing on memes that are necessary to decode their meaning, but also the contextualising web of meanings defined by the accompanying hashtags. Analysing hashtags reveals how users connect the purpose of the post to other popular points in conspiracy theories. Thus, for example,

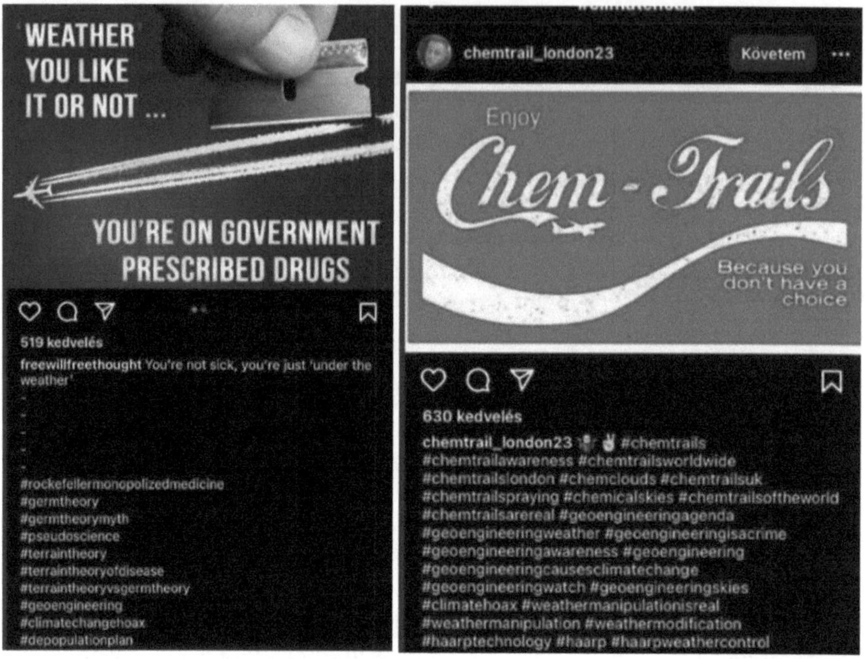

Figure 3.5 Left: [@freewillfreethought] (2022, December 11). *You are not sick, and you're under the weather.* Meme][Screenshot by author]. Instagram. www.instagram.com/p/CmAe0-INb4S/Right: [@chemtrail_london23] (2023, March 19). *#chemtrails #chemtrailawarness #chemtrailsworldwide.* Meme][Screenshot by author]. Instagram. www.instagram.com/p/Cp-nMEdtqA7/

the meme that primarily denotes fears about technology describes not only 5g-related conteos but points to climate denial, mandatory bug eating as part of an environmentalist agenda, the communist world conspiracy and even the plan to exterminate humanity (see Figure 3.6.).

The sample is characterised by a robust negative referentiality, which stems from the fact that although conspiracy theories promote an ideology and world explanation, they primarily take a countering position. They oppose scientific facts and the social, political and economic institutions. In the case of climate change, this means, in addition to the scientific position, they oppose and query the whole scientific society, the environmentalists and politicians and other actors of the establishment making climate-related laws and enforcing these.

The role of the media is also controversial among the deniers because although at the level of denotation, hashtags only make the media appear in the part of the conspirator *(#mediagenda, #mediacorruption, #mediapropaganda, #fakenews*), a high percentage (10%) of the posts contain news from media organisations, which themselves are appearing on a platform of (social) media. In this context, it is essential to note that the snippets taken from the news media are often print screens

Figure 3.6 [jesus_murphy306]. (2023, March 4.)#digitalslavery #eatthebugs #worldgovernment [Meme][Screenshot by author]. Instagram. www.instagram.com/p/CpYI78AOsYS/

of texts, which reinforces the verbal focus of climate-denier content on Instagram. The source of news shows a high diversity, with CNN and CNBC appearing as well as (far)-right outlets such as Fox News, Breitbart or PragerU. However, the former are not presented as fake news but are used to justify the exposure of various agendas.

The issue of masks and lockdowns, i.e. linking climate denial to COVID-19, is an excellent example of how social media pseudo-news and conspiracy theories incorporate current phenomena. This connotation begins with the demonisation of the green agenda, fear-mongering against climate change policies and culminates in the dystopia of a climate change lockdown (*#climatelockdown)*. The coronavirus epidemic was accompanied by a proliferation of conspiracy theories (Bierwiaczonek, Gundersen, & Kunst, 2022) that formulated and disseminated theories about the origin of the coronavirus (*#plandemic*), control policies and vaccines. The list of conspiracy theories about SARS-CoV-2 (Fuchs, 2021) includes that Israel, the Jews, the Rothschild family or George Soros developed the virus to take over the world. Among these, George Soros, the Rothschilds and the Rockefellers are also repeatedly mentioned in posts tagged #climatehoax. Bill Gates is one of the central figures of conspiracy theories on COVID-19 and the most frequently cited person on the level of hashtags in the sample analysed. There is also an overlap in the fear-mongering around 5G, which is blamed for the COVID-19 outbreak and occurs in 4% of the hashtags in the corpus. According to COVID-related conspiracies, the deliberate manufacturing of the virus aimed to create a new world order in which people everywhere are monitored and tracked with microchips (Fuchs, 2021). Although microchips rarely appear in the content analysed, the new world order is among the frequently occurring hashtags and is regularly mentioned with other prominent hashtags such as population agenda, globalist agenda, and conspiracy theory. The latter is used both ironically and seriously, i.e. it is both an awareness of their stigmatisation and ironic rejection of it and a form of self-stigmatisation, signalling identification with the group of conspiracy theory believers. From a discursive point of view, fear-mongering, in-group victimisation and out-group demonisation characterise this set of conspiracy theories. The corpus is imbued with a sense of outsiderness and excellence, indicated by the rhetorical terms *wake up*, *awaken*, *when will you wake up* and its *sheep* counterpart. Claiming oneself as awakened is a common mode of identification in the narratives of conspiracy theories. There is even a group calling themselves We Are Awakening (Harambam, 2020). Barkun (2013) distinguishes three types of conspiracy theories: systemic, event-related, and super. Climate denial is a systemic conspiracy theory associated with conspiracy theories about COVID-19, 5G, smartmeters, Bill Gates, government intrigue, leftist evil, malicious pharmaceutical companies, chemtrails, and the rejection of the sphericity of the Earth. These systemic conspiracy theories are randomly combined along the hashtags forming an incomplete and self-contradictory super conspiracy theory whereby the left and the governments of the day are harming people with chemicals and viruses to create a new world order, and the scientists paid by the depopulation agenda are spreading lies about anthropogenic climate change and spherical Earth. The conspiracy is motivated by money and power, but the theory does not address how the extermination of humans would serve these ends.

Articulating climate denial visually on Instagram

Since this research aimed to explore the visual language of climate denial communication as part of the discourse on Instagram, I will focus on the posts that visually articulate climate denial in this part of the analysis. As an analogy to the most iconic object of climate change communication, one could ask: What's their polar bear like? Does this group even have its iconography, and if so, how does it appear on Instagram, one of the least researched but prevalent social media platforms?

As aforementioned, 77% of the corpus is only textually related to climate denial; thus, less than a fourth of the canon does visually formulate the issue, which might indicate that denying this global phenomenon visually is just as tricky as claiming its existence. Climate-denier iconography in the analysed sample tends to reuse and rewrite the imagery used by climate-positive visual communication. The majority of visual climate change denialist posts use existing social media (8%) and news media (15%) images or rewrite internet memes (27%). Research on climate-positive imagery found that media uses a small number of sources, primarily images from commercial image agencies (O'Neill, 2017). Climate change results in a well-defined iconography, e.g., melting glaciers, polar bears, globe, smoke stacks, deforestation, melting ice, glaciers etc.(O'Neill, 2017, Wang et al., 2018). It is also indicated that a common argument among climate sceptics is to deny the existence of climate change by referring to actual weather events such as snow storms occurring in seasons or locations where it's supposed to be warmer (Lawrence & Estow, 2017; Pearce et al., 2019). Accordingly, a proportion (8%) of visual-climate denier messages in the sample analysed were also pictorially based on weather phenomena. From this data, it can be concluded that, on the one hand, climate denial on Instagram is articulated through textual rather than visual messages. On the other hand, negative referentiality is also manifest at the level of the climate denial imagery in the corpus.

Climate sceptics re-frame the climate change-related visuals used and established by the media according to their narrative. Accordingly, the topos and personalities of a mediatised, global iconography of climate change re-appear in the visual communication denying climate change, which is not surprising given that this communication itself takes place on a global social media platform. This is illustrated by the fact that – with the exception of two posts showing Al Gore – posts that depict climate denial visually with a public figure (4% of the whole corpus) feature Greta Thunberg. The Swedish activist who has become an iconic figure since the 2019 climate protests is being viciously attacked by climate deniers on Instagram, not only visually by committing virtual violence by disfiguring her face but also verbally. Her persona is identified with climate change as a whole. That is, by portraying her as an incompetent and ridiculous child, the climate change phenomenon is ridiculed. The rhetorical device of impersonation seems to help visualise the otherwise abstract climate change phenomenon. Thunberg's face often appears distorted (e.g. as Gollum or as an old lady or blended with the face of right-wing political influencer Ben Shapiro, etc.) and embedded in a meme. Meme versions of her often make fun of her speech which she held in 2019 at U.N.'s Climate Action Summit in New York City, in which she repeatedly asked world leaders, "How dare you?" i.e. take away the future of generations for the sake of profit. Her anger, young age and solid facial expression are being ridiculed as she is depicted as a character of Game of Thrones, a pop cultural reference and, at the same time, part of climate change's iconography (i.e. weather events, snow).

Memes recycle, parody and paraphrase elements of visual mass culture. They are also characterised by easy reproducibility, humour and, from an aesthetic point of view, a lack of sophistication (Zhang & Pinto, 2021; Shifman, 2013). In the latter context, it is essential to note that the corpus uses language which differs from the platform vernacular of Instagram in every aspect. There is no documentary value to the images; they are not aestheticising and do not refer to the "now" but convey knowledge – albeit false knowledge. Their communication is not about aesthetic pleasure or capturing a moment but conveying information and persuasion. Aesthetics is wholly subordinated to their rhetorical function, so they often work with memes or texts superimposed on posts.

Despite their apparent lightness and humour, memes often express criticism, concern and messages aimed at challenging the hegemony of the status quo. The subversive quality of memes is also prominent in the corpus under study as they critique the establishment in this context (see Figure 3.7.). Conspiracy theories born digitally often expand from one issue to another, such as the theory of the Great Reset, often used as a tag in the corpus. The latter approach initially alleged that the evil elite deliberately engineered COVID-19, but later on, it expanded to climate change, inflation related to renewables, and the economic measurements taken against Russia since its war on Ukraine, etc. The second print screen of Figure 3.7. illustrates how a single post can communicate central ideological, public policy and economic guidelines and preferences with a well-known meme character. This post ties together conspiracy theories related to COVID-19, climate change, the war in Ukraine and inflation; simultaneously, it opposes leftist ideology. The latter is visually formulated through a mohawk hairstyle associated with the punks belonging politically to the far left, whereas its caption reads "Climate-socialism!". The anti-leftist message is made even more explicit by hashtags such as *#liberalismisamentaldisorder*, and the political affiliation of the uploader is made more concrete by hashtags such as *#makeamericagreatagain* and *#trumpismypresident*. The figure with glasses blinded by the Ukrainian flag, with the caption "War with Russia!" is a clear statement in favour of the Russians. It reinforces the official Russian narrative that the West is at war with Russia and that they are not the aggressors against Ukraine. This post functions as both a world explanation and an ideological creed in the meme section and also contains a clear party-political statement at the hashtag level. In addition, the uploader who takes a pro-Trump and strongly anti-Biden and anti-Democrat stance (i.e. *#bidensucks #democratsdestroyameirca*) identifies himself with the hashtags #blackconservative and *#blacksfortrump*. The echoing of climate denialist, anti-vaccination, and anti-mask-wearing, pro-Russian narratives by a Trumpist user illustrates the radicalisation of mainstream right-wing politics and the prevalence of previously marginalised conspiracy theories along populist narratives.

Conclusion

Climate denial communication on Instagram is characterised by negative reflexivity that embeds scepticism about anthropogenic climate change in a broad web

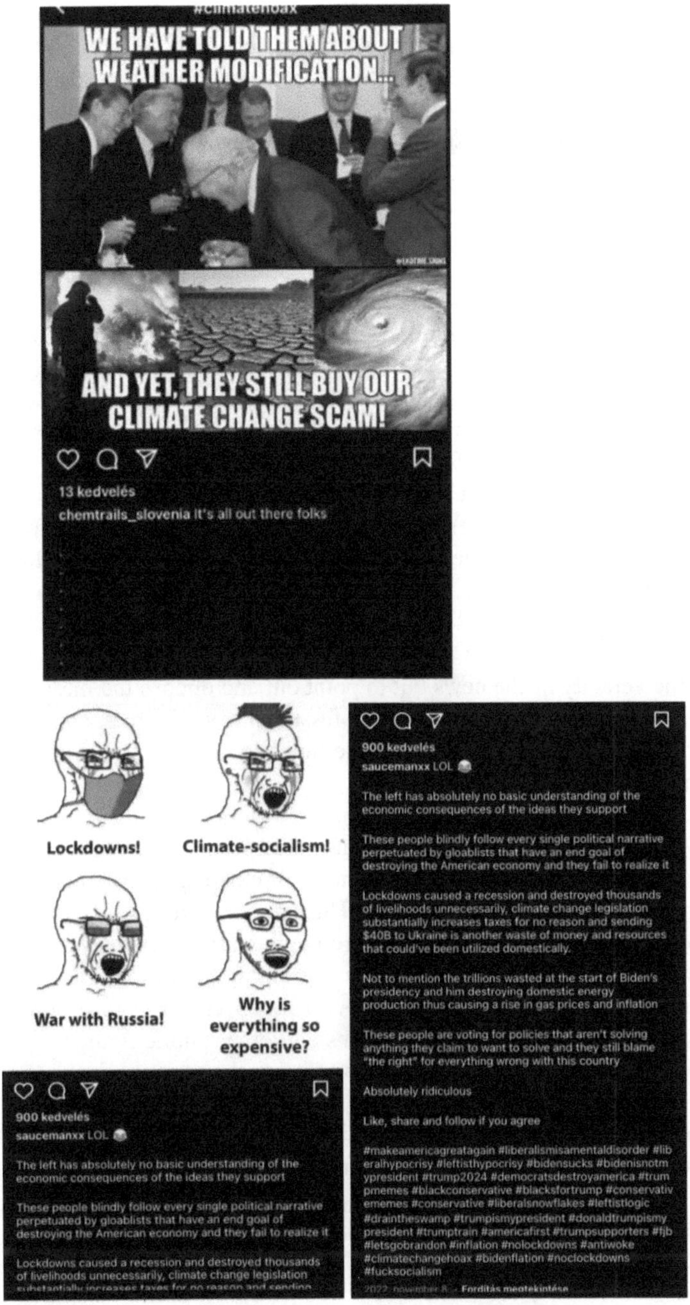

Figure 3.7 From left to right: [@chemtrails_slovenia] (2022, November 15). *It's all out there, folks*. Meme][Screenshot by author]. Instagram. www.instagram.com/p/Ck_Lx6SNU4E/ [@saucemanxx] (2022, November 8). *LOL*. Meme][Screenshot by author]. Instagram. www.instagram.com/p/Ckrn_pxNIL2/

of conspiracy theories. Among these, science-denying flat Earth theories and the chemtrail theory linked to the depopulation agenda are dominant and have already defined the spread of misinformation in relation to COVID-19 (Fuchs, 2021). These logically flawed conspiracy theories, linked to far-right populist ideology, culminate in a super conspiracy theory, through the use of hashtags. In contrast to previous research on climate scepticism on Twitter, the content here is not event-driven but abstract and systemic.

Climate sceptic content on Instagram is unlike the platform vernacular in that it is not aestheticising and does not aim to capture a moment in time. Instead, it is text-centric and aims to be informative – be it fake information. Texts appear as part of memes, as re-shared news from media outlets and Twitter posts, in the form of image-less textual posts, captions, hashtags, and images. The extensive use of the latter often gives the base a surplus meaning beyond the visual content, placing it in an ideological framework. These groups of hashtags point to the terrain of systemic conspiracy theories. The platform's internal logic indirectly helps promote these as hashtag aggregation is done to achieve greater reach and popularity. The verbal nature of climate denial communication on Instagram (77% of the corpus is only textually related to climate denial) is reinforced by the re-sharing content from news media sites (10% of the sample) and other social media sites (7%) which draw on a variety of sources. Besides news organisations representing right-wing ideology, such as Fox News and Breitbart, other more objective news sources are also used (e.g. CNN, BBC). The latter's content is redistributed in a revealing way, not to challenge the veracity of the news but to point out and oppose the measures they describe. Despite the frequent citation of professional news sources, concerns about traditional media losing its gatekeeper role must be acknowledged, as laic users generate the vast majority of content.

Results show that climate denial is hardly visualised (less than a quarter of the corpus). Still, if it is, it is rarely through original imagery but rather through the distortion or re-framing of particular visual elements of existing climate communication accompanied by textual information. Thus negative reflexivity is also manifest at the level of visual communication through the large-scale recycling of images. Image recycling includes memes and the re-framing of familiar images of climate change in the news media, such as depictions of weather events, photos of sea level rise and arctic ice melting, politicians and other public figures, most prominently Greta Thunberg. The Swedish activist has become an impersonator of climate change in climate-positive as well as climate denial communication. Functioning as a condensed symbol, in the latter, her gestures, face and messages are viciously attacked and ridiculed in the form of humorous memes. The imagery and the hashtags applied also indicate that climate deniers are using a globally usable, sufficiently general and accessible language, compatible with social media, which recycles mainstream media's climate communication to suit their own purposes.

Humour, irony and mediatisation are now the primary weapons of radical right-wing movements in communication. According to Jenkins et al. (2013), humour and parody are among the most critical factors that help a given content to spread in a networked culture because they create a community among those who understand the joke. This light-hearted tone marks a significant change from using national socialist ideological symbols which presupposed historical knowledge. This use of mass culture as a message carrier, a medium and a point of reference has made far-right movements and groups dominant actors in internet culture.

Further research exploring the visual communication used by climate sceptics on different social media and news media platforms could shed light on the reasons for the presence of image recycling as well as add more detail on the source of this type of misinformation.

Notes

1 13,158 posts were found using the #climatechangehoax and 8633 using the hashtag climatehoax on 17.04.2023.

2 It is important to note that the latter category includes pots by an NFT artist, a self-proclaimed Dutch artist who has flooded one half of his sample (i.e.: the one tagged #climatehoax). The images uploaded by him, which on average receive 20–35 likes are portraits with infantile aesthetics. With one exception, they do not address climate change in their visual content or description, and the hashtags also only include the one: #climatehoax in relation to the topic of climate denial. The one exception mentioned above makes it clear however that the use of the climate denial tag is not applied by mistake or misunderstanding. Despite their overwhelming number, posts by the NFT artist are not a relevant part of the sample in terms of substance, which is why I did not include them in the research any further than this step as they were predominantly related to the art sphere and self-promotion.

3 A term I use to address communication which conforms to the reality of anthropogenic climate change.

References

Allgeier, J. (2019). Science and environmental communication on youtube: strategically distorted communications in online videos on climate change and climate engineering. *Frontiers in Communication.Sec. Science and Environmental Communication.* https://doi.org/10.3389/fcomm.2019.00036

Anderson, A.A. (2017). Effects of social media use on climate change opinion, knowledge, and behavior. In H. von Storch (Ed.), *Oxford research encyclopedia of climate science.* Oxford. https://doi.org/10.1093/acrefore/9780190228620.013.369

Barkemeyer, R., Figge, F., Hoepner, A., Holt, D., Kraak, J. M., & Yu, P.-S. (2017). Media coverage of climate change: An international comparison. *Environment and Planning C: Politics and Space, 35*(6), 1029–1054. https://doi.org/10.1177/0263774X16680818

Barkun, M. (2013). *A culture of conspiracy: Apocaliptyc visions in contemporary America.* University of California Press.

Bierwiaczonek, K., Gundersen, A.B., & Kunst, J.R. (2022). The role of conspiracy beliefs for COVID-19 health responses: a meta-analysis. *Current Opinion in Psychology, 46,* 101346. https://doi.org/10.1016/j.copsyc.2022.101346

Björnberg, K.E., Karlsson, M., Gilek, M., & Hansson, S.O. (2017). Climate and environmental science denial: a review of the scientific literature published in 1990–2015. *Journal of Cleaner Production, 167*, 229–241. https://doi.org/10.1016/j.jclepro.2017.08.066

Boussalis, C., & Coan, T.G. (2016). Text-mining the signals of climate change doubt. *Global Environmental Change, 36*, 89–100. https://doi.org/10.1016/j.gloenvcha.2015.12.001.

Burgess, J. (2006). Hearing ordinary voices: cultural studies, vernacular creativity and digital storytelling. *Continuum: Journal of Media & Cultural Studies, 20*(2). 201–214. https://doi.org/10.1080/10304310600641737

Butter, M. (2022). Conspiracy theory after trump. *Social Research: An International Quarterly, 89(*3), 787–809.https://doi.org/10.1353/sor.2022.0054.

Butter, M., & Knight, P. (Eds.). (2020). *Routledge handbook of conspiracy theories*. Routledge. https://doi.org/10.4324/9780429452734

Cairns, R. (2016). Climates of suspicion: "chemtrail" conspiracy narratives and the international politics of geoengineering. *The Geographical Journal, 182*(1), 70–84. www.jstor.org/stable/43868685

Carvalho, A. (2007). Ideological cultures and media discourses on scientific knowledge: re-reading news on climate change. *Public Understanding of Science, 16(*2), 223–234. https://doi.org/10.1177/0963662506066775

Cook, J., Oreskes, N., Doran, P.T., Anderegg, W.R., Verheggen, B., Maibach, E.W.,... & Rice, K. (2016). Consensus on consensus: a synthesis of consensus estimates on human-caused global warming. *Environmental Research Letters, 11*(4), https://doi.org/10.1088/1748-9326/11/4/048002

Dixon, S. (2023, February 15). Instagram: number of global users 2020-2025. Statista. www.statista.com/statistics/183585/instagram-number-of-global-users/

Domingo, M., Jewitt, C., & Kress, G. (2020). Multimodal social semiotics: writing in online contexts. In J. Roswell & K. Pahl (Eds.), The Routledge Handbook of Literacy Studies. Routledge.

Dotson, D.M., Jacobson, S. K., Kaid, L.L. & Carlton, J.S. (2012). Media coverage of climate change in chile: a content analysis of conservative and liberal newspapers. *Environmental Communication, 6*(1), 64–81. https://doi.org/10.1080/17524032.2011.642078

Douglas, K.M., Uscinski, J.E., Sutton, R.M., Cichocka, A., Nefes, T., Ang, C.S., & Deravi, F. (2019). Understanding conspiracy theories. *Political Psychology, 40*, 3–35. https://doi.org/10.1111/pops.12568

Dunlap, R.E., McCright A.M. & Yarosh, J.H. (2016). The Political divide on climate change: partisan polarization widens in the U.S. *Environment: Science and Policy for Sustainable Development, 58*(5), 4–23, https://doi.org/10.1080/00139157.2016.1208995

Ekman, M. (2022). The great replacement: strategic mainstreaming of far-right conspiracy claims. *Convergence, 28*(4), 1127–1143. https://doi.org/10.1177/13548565221091983

Elsasser, S.W., & Dunlap, R.E. (2013). Leading voices in the denier choir: conservative columnists' dismissal of global warming and denigration of climate science. *American Behavioral Scientist, 57*(6), 754–776. https://doi.org/10.1177/0002764212469800

Fairclough, N. (2015[1989]). *Language and power,* 3rd edition.Routledge.

Forchtner, B. (2019). Climate change and the far right. *Wiley Interdisciplinary Reviews: Climate Change, 10*(5), e604. https://doi.org/10.1002/wcc.604

Fuchs, C. (2021). *Communicating COVID-19: everyday life, digital capitalism, and conspiracy theories in pandemic times*. Emerald Group Publishing.

Gomes, M. & Böhm, S. (2022). Right-wing populism versus climate capitalism: Climate change governance under scrutiny. In G. Morgan & M. Feldmann (Eds). *Business elites and populism*, Oxford University Press.

Han, J., Sun, S. & Lu, Y. (2017). Framing climate change: a content analysis of Chinese mainstream newspapers from 2005 to 2015. *International Journal of Communication, 11*, 2889–2911

Harambam, J. (2020). *Contemporary conspiracy culture: truth and knowledge in an era of epistemic instability.* Routledge.

Herrmann, C., Rhein, S., & Dorsch, I. (2022). # fridaysforfuture–What does instagram tell us about a social movement?. *Journal of Information Science,* https://doi.org/0165551521 1063620

Highfield, T., & Leaver, T. (2016). Instagrammatics and digital methods: studying visual socialmedia, from selfies and GIFs to memes and emoji.*Communication Research and Practice,2*(1), 47–62. https://doi.org/10.1080/22041451.2016.1155332

Howard P.N., Ganesh, B., & Liotsiou, D. (2019*). The IRA, social media and political polarization in the United States, 2012–2018.* Computational Propaganda Project, Oxford University. Available at: https://demtech.oii.ox.ac.uk/wp-content/uploads/sites/12/2018/12/The-IRA-Social-Media-and-Political-Polarization.pdf

Huber, R.A. (2020). The role of populist attitudes in explaining climate change skepticism and support for environmental protection, *Environmental Politics*, *29*(6), 959–982, DOI: https://doi.org/10.1080/09644016.2019.1708186

Jenkins, H., Ford, S., & Green, J. (2013). *Spreadable media.* New York University Press. www.jstor.org/stable/j.ctt9qfk6w

Kulin, J., Sevä J. & Dunlap, R.E. (2021). Nationalist ideology, rightwing populism, and public views about climate change in Europe. *Environmental Politics, 30*(7), 1111–1134, https://doi.org/10.1080/09644016.2021.1898879

Lawrence, E.K. & Estow, S. (2017) Responding to misinformation about climate change, *Applied Environmental Education & Communication, 16(*2), 117–128, https://doi.org/10.1080/1533015X.2017.1305920

Lockwood, M. (2018). Right-wing populism and the climate change agenda: exploring the linkages. *Environmental Politics*, *27*(4), 712–732, https://doi.org/10.1080/09644016.2018.1458411

Machin, D., & Mayr, A. (2023). *How to do critical discourse analysis: a multimodal introduction*. Sage.

McCright, A.M., & Dunlap, R.E. (2011). The politicization of climate change and polarization in the American public's views of global warming, 2001–2010. *The Sociological Quarterly*, *52*(2), 155–194. https://doi.org/10.1111/j.1533-8525.2011.01198.x

McCright, A.M., Dunlap, R.E., & Marquart-Pyatt, S.T. (2016). Political ideology and views about climate change in the European Union. *Environmental Politics, 25*(2), 338–358. https://doi.org/10.1080/09644016.2015.1090371

Newman, T. P. (2017). Tracking the release of IPCC AR5 on twitter: users, comments, and sources following the release of the working Group I summary for policymakers. *Public Understanding of Science, 26*(7), 815–825. https://doi.org/10.1177/0963662516628477

O'Neill, S.J. (2017). Engaging with climate change imagery. In H. von Storch (Ed.), *Oxford encyclopedia of climate change communication.*

Painter, J., & Gavin, N. T. (2015). Climate skepticism in British newspapers, 2007–2011. *Environmental Communication, 10*(4), 432–452. hhttps://10.1080/17524032.2014.995193.

Pearce, W., Niederer, S., Özkula, S.M., & Sánchez Querubín, N. (2019). The social media life of climate change: platforms, publics, and future imaginaries. *Wiley Interdisciplinary Reviews: Climate Change, 10*(2), e569. https://doi.org/10.1002/wcc.569

Poortinga, W., et al., 2019. Climate change perceptions and their individual-level determinants: a cross-European analysis. *Global Environmental Change, 55*, 25–35. https://doi.org/10.1016/j.gloenvcha.2019.01.007

Rieder, B., Matamoros-Fernández, A. & Coromina, O. (2017). From ranking algorithms to "ranking cultures": investigating the modulation of visibility in youtube search results. *Convergence: The International Journal into New Media Technologies, 24*(1), 50–68. https://doi.org/10.1177/1354856517736982

Rogers, R. (2021). Visual media analysis for Instagram and other online platforms. *Big Data & Society, 8*(1), https://doi.org/10.1177/20539517211022370

Shifman, L. (2013). Memes in a digital world: reconciling with a conceptual troublemaker, *Journal of Computer-Mediated Communication, 18*(3), 362–377, https://doi.org/10.1111/jcc4.12013

Treen, K.M.D.I., Williams, H.T., & O'Neill, S.J. (2020). Online misinformation about climate change. *Wiley Interdisciplinary Reviews: Climate Change, 11*(5). https://doi.org/10.1002/wcc.665

van der Linden, S., Leiserowitz, A., Rosenthal, S., & Maibach, E. (2017). Inoculating the public against misinformation about climate change. *Global Challenges, 1*(2), https://doi.org/10.1002/gch2.201600008.

van Dijk, T. (2002). Multidisciplinary CDA: a plea for diversity. In R. Wodak & M. Meyer (Eds.), *Methods of critical discourse analysis* (pp. 95–121). Sage.

Vraga, E.K., Kim, S.C., Cook, J., & Bode, L. (2020). Testing the effectiveness of correction placement and type on instagram. *The International Journal of Press/Politics, 25*(4), 632–652. https://doi.org/10.1177/1940161220919082

Wall, J.D., Stahl, B.C., & Salam, A.F. (2015). Critical discourse analysis as a review methodology: an empirical example. *Communication Association for Information Systems, 37*, 11. https://doi.org/10.17705/1CAIS.03711

Wang, S., Corner, A., Chapman, D., & Markowitz, E. (2018). Public engagement with climate imagery in a changing digital landscape. *Wiley Interdisciplinary Reviews: Climate Change, 9(*2). https://doi.org/10.1002/wcc.509

Williams, H.T., McMurray, J.R., Kurz, T., & Lambert, F.H. (2015). Network analysis reveals open forums and echo chambers in social media discussions of climate change. *Global Environmental Change, 32,* 126–138. https://doi.org/10.1016/j.gloenvcha.2015.03.006

Wu, L., Morstatter, F., Hu, X., & Liu, H. (2016). Mining misinformation in social media. In M.T. Thai, W. Wu & H. Xiong (Eds.), *Big data in complex and social networks* (pp. 135–162). Routledge.

Zhang, B., & Pinto, J. (2021). Changing the world one meme at a time: the effects of climate change memes on civic engagement intentions. *Environmental Communication, 15*(6), 749–764. https://doi.org/10.1080/17524032.2021.1894197

4 Cli-fi and five narratives of future warming

Gregers Andersen

Introduction

In the fourth of the six Gilford Lectures that the late Bruno Latour gave at the University of Edinburgh in February 2013, one finds an important reflection on the role of aesthetics at the threshold of the Anthropocene: the geological epoch that has come to epitomize the destructive geophysical imprints of human activity. In the dying minutes of the lecture, titled *The Anthropocene and the Destruction of The Image of the Globe*, Latour claimed that in the wake of the disastrous future that the Anthropocene may entail, it is "our ideal ideas of the globe that should be destroyed for any work of art, any aesthetic to emerge, if you agree to hear in the word aesthetics its old meaning of being able to perceive and to be concerned. That is a capacity to render oneself sensitive" (Latour 2013). Latour did in this way emphasize that the (eco)modernist project of turning Earth into a globalized network of economic exchange and transactions (i.e. a global market) has reached an impasse, as the desires, logics, and representations driving this project appears to be at odds with the forms of sensibilities required for humans to reconnect with the Earth System and become sustainable beings.

In what follows I will take Latour's remarks as an invitation to delve deeper into one of the new forms of aesthetics that has emerged with this situation. More precisely, the departure point of my inquiry will be the research question: How have literary and cinematic fiction from the Global North (i.e. North America, Australia, and Western Europe) engaged with the reality of rapidly accelerating anthropogenic global warming? By selecting this research question I aspire not only to shed more light on some of the literary and cinematic fiction that have during the last five decades emerged as an aesthetic response to the human warming of the planet. I intend also to demonstrate how these fiction represent important affective, cognitive, and imaginative supplements to the climate scientific projections appearing in the reports by UN's Intergovernmental Panel on Climate Change (IPCC) as well as in other scientific publications. In this regard, I take peer-reviewed projections of different scenarios of future climate change to be very important and trustworthy constructions of truth. But at the same time, they are also quite unsatisfying in the sense that there are clear limits to what they include. Although the projections are constructs of futures that reality could imitate (with different degrees of

DOI: 10.4324/9781003479550-5

likelihood) there are clearly forms of speculation that the projections do not engage in due to scientific norms. This includes, for example, detailed speculations about how humans might affectively and cognitively experience the worlds that the projections sketch. It includes detailed speculations about which societies these worlds could engender and how they might handle increasingly scarce resources, growing streams of refugees or more sensibly care for vulnerable ecosystems – to give just a few examples. The point being that the scientific projections leave this kind of speculation to fictive descriptions of anthropogenic global warming, that is, to what is now more commonly known as climate fiction.

The term climate fiction – and its catchier abbreviation: cli-fi – was coined by American blogger Danny Bloom in 2007. Since then the term has come to define fiction of various genres that employ the scientific paradigm of anthropogenic global warming in their plots and world-making (Andersen 2016; Johns-Putra 2016). Defined in this way cli-fi simply means fiction that "uses as a narrative element the scientific consensus that humanity's emissions of greenhouse gasses cause global warming" and in the Global North has a history that goes back to at least the early 1970s (Andersen 2020: 5). For example, in the early American 1970s film *Soylent Green* (1973) the greenhouse effect is referred to as having caused the heated and overpopulated world that the film depicts. This is, however, not a connection that is unfolded in detail. It serves strictly as an explanation for the setting and not as a vehicle for further reflection or action. In this sense, cli-fi is more properly born in the Global North a few years later with the publication of the thriller *Heat* (1977) by American author Arthur Herzog. In Herzog's thriller humanity's emissions of greenhouse gasses are not only represented as the cause of "runaway" warming, but also framed and reflected upon as a problem fundamental to human existence (1989:270). In the short format of this chapter, I am unable to offer a thorough analysis of all literary and cinematic cli-fi that from *Heat* and onwards have been produced and published in the Global North. Instead, I will seek to answer my research question by introducing five narrative templates that have dominated in cli-fi from the Global North. When I claim that these five narrative templates have 'dominated' the genre, it is because they can be identified in a broad sample of climate fiction. My identification of these five templates is thus the outcome of a comparative study of more than hundred different literary and cinematic works of cli-fi from the Global North. Methodologically, these works have been arbitrarily selected with the only common denominators being that they mainly stem from North America, Australia, and Western Europe (including Great Britain) and therefore are chiefly (but not exclusively) anglophone. Consequently, one might say that the five narrative templates represent structures or patterns that run through cli-fi from the Global North. However, such a formulation should not be made without critical reflection, as any reference to terms such as structures or patterns immediately begs the question as to whom these structures or patterns are supposed to belong. Thus, to be more precise, I perceive the narrative templates that this article will map as being intrinsic to the climate fiction they are part of, in the sense that their existence can be inductively demonstrated. Yet they are also categories of interpretation rather arbitrarily termed. Key in this regard is that the templates work as templates for world-making, that is, they function as cognitive schemas that confine the artistic imagination to certain starting points, but still open it to the creation of a multitude of possible worlds.

A fundamental question is therefore also how the existence of this repertoire of narratives can be explained in more philosophical and epistemological terms. In this regard, I believe a satisfying answer can be found in hermeneutical philosophy; and what Hans-Georg Gadamer referred to as its "ontological orientation" (2006: 296). The core of this orientation is to be found in the work of Gadamer's mentor, Martin Heidegger, namely in paragraph 31 and 32 of his magnum opus: *Being and Time* (1927). Here, Heidegger not only ontologically establishes how understanding is to be viewed as a fundamental condition for human existence. He also points to the "fore-structure of understanding", which means that human subjects are always already part of a world (i.e. cultural context), shaping their understanding (2001:192). This explains, to a certain extent, the narrative structures that I have identified (via comparison) in my sample of cli-fi from the Global North. But blending Heidegger's argument with one of the classical arguments of cultural anthropology it is possible to stretch this argument even further. Hence if culture is indeed, as the Canadian anthropologist Clifford Geertz once claimed (1973), "the historically created systems of meanings in term of which we give form, order, point, and direction to our lives", one could with inspiration from Heidegger add that these systems not only pre-structure understanding (52). They should also be thought of as something that continuously feeds the imagination by constituting its premises and foundations.

In fact, French anthropologist Claude Lévi-Strauss described something similar with the term "bricolage" (2004:16). In *The Savage Mind* (1962) Lévi-Strauss sets out to demonstrate how the mythmaker saves the world picture of his culture by rearranging fragments of different myths every time this world picture is challenged or put in danger by new, unforeseen events. Equivalent to building blocks which do not only stay the same, but also give form to the new brickwork that continuously cover them, the five narratives I will present contain some of the same qualities. However, my presentation of the five narratives relies on an observation that both follows and goes somewhat against Lévi-Strauss's description of how "mythical thought builds up structures by fitting together remains of events, while science creates its means and results in the form of events" (22). Thus, instead of confirming the existence of the gap that Lévi-Strauss found to divide the mode of bricolage from the mode of science, I believe it is more fruitful to see these two modes as intertwined. To be more precise, I perceive bricolage as a natural cognitive response to the events and the paradigmatic shifts produced by science. Consequently, I will not only introduce the five narratives I have found to dominate in cli-fi from the Global North. I will also point to relations between these narratives and some of the ancient myths that have played a key role in the cultural history of the West.

But first a word of caution. My focus on cli-fi produced in Europe, Australia, and North America is obviously guilty of reproducing a problem that has haunted cli-fi research for more than a decade. Although several monographs and anthologies on cli-fi have appeared in the last ten years, an overwhelming majority of the research has been on literary cli-fi written by authors residing in either Great Britain or North America (e.g., Trexler 2015; Clark 2015; Mehnert 2016; Bracke 2018; Goodbody & Johns-Putra 2018; Johns-Putra 2019a; Johns-Putra 2019b; Andersen 2020). Consequently, the field of cli-fi research lacks perspectives on cli-fi from other nations and cultural spheres. Particularly there is a deficit of research on cli-fi coming from the Global South. This represents a major problem. Not only because the escalation of the Anthropocene causes different existential and societal situations around the planet, but more importantly also because the escalation of climatic and ecological disasters particularly threatens the Global South. My focus below does not amend this problem. But I hope that my mapping of key narrative templates in cli-fi from the Global North can at least provide a basis for still broader explorations in the field. So let me now turn toward the five templates that I have found to dominate in cli-fi from the Global North in the past five decades.

The social collapse: War and post-apocalyptic violence

Shaping the world-making in a broad spectrum of North American, European, and Australian cli-fi one finds the narrative template of "The Social Collapse". According to this template (or imaginary) anthropogenic global warming will lead to social disintegration and interhuman violence on a very large scale. On a macro level the consequence of this disintegration could be war as in Matthew Glass's novel *Ultimatum* (2009), where a geopolitical feud over the reduction of greenhouse gasses ends in nuclear war. However, much more commonly the disintegration appears on a micro level in the form of individuals battling for limited resources in a post-apocalyptic world where the social contract has ruptured. Among the climate fiction unfolding such a scenario are Steven Amsterdam's novel *Things We Didn't See Coming* (2009) as well as Martine McDonagh's novel *I Have Waited, And You Have Come* (2012). However, it is Marcel Theroux's post-apocalyptic novel *Far North* (2009) and Jeff Renfroe's post-apocalyptic cli-fi movie *The Colony* (2013) I will here momentarily turn to, because in both fiction we find a quite obvious connection between the template of The Social Collapse and the biblical tale of Babel. By this I do not mean that these two fiction display a punishing God that effectively puts a presumptuous humanity back into its place by dividing it up into antagonistic clans. The connection is more subtle, as in both climate fiction humanity's civilizational progress is disrupted and reversed by humanity's own scientific and technological aptitude. First, industrial modernity leads to anthropogenic global warming, then sets into motion a world wherein climate refugees incapable of communicating are engaged in a battle to death. Furthermore, this regressive evolution is in both fiction symbolized by architecture. In *Far North*, the main character Makepeace traces some of the devastation that has struck the post-apocalyptic world to the city of Polyn: a city that "by its size and wealth might as well have been built by gods as men", but in its present post-apocalyptic state only makes "a mockery" of Makepeace's "patched cloth and scavenged food" (Theroux, p. 191). In other words: Polyn encapsulates a fatal promethean ambition.

It is the symbol of a humanity that has striven toward the skies, but in its endeavor brought destruction upon itself. This Babel-like theme is integrated further into the plot by a general breakdown in interhuman communication. In the devastated Siberian woodlands Makepeace once regarded as a safe haven the general collapse in human civilization is mirrored by a communicative turn from language to violence. As different generations of climate refugees incapable of speaking to each other clash over scarce resources, verbal expression gives way to raw physical expressions.

A similar plot is unfolded in *The Colony*. However, the fatal promethean ambition is here encapsulated in a number of large "weather modification towers" that "did their damage a long time ago" (Renfroe 2013). These towers stand as slowly decaying monuments above a frozen world almost entirely bereft of life. The viewer is told at the beginning of the film how humanity had tried to manage the threat of accelerating anthropogenic global warming by using the towers to modify the weather many years earlier. However, even though the costs of this attempted mastery has been extremely high (the few humans that still exist are forced to live underground) the beginning of the film still introduces a micro-world of relative harmony. Order to some extent still exists in the colony at the center of the film's focal point, even though its maintenance is displayed as both difficult and painful. But as in so many other post-apocalyptic fiction this is only the calm before the storm. After the colony sends out an expedition to help another colony, its inhabitants soon find themselves in a battle for life and death with a horde of bloodthirsty cannibals. These cannibals represent an admittedly extreme, but nonetheless quite logical low point to the negative social anthropology that generally appear in conjunction with the narrative template of The Social Collapse. The cannibals depicted in *The Colony* thus mark the culmination of the violent and barbaric human beings that are not only dominant in climate fiction utilizing the narrative template, but also feature in a number of popular scientific books such as Gwynne Dyer's *Climate Wars* (2008), Christian Parenti's *Tropic of Chaos* (2011), and Harald Welzer's *Climate Wars: What People Will be Killed for in the 21st Century* (2012). Thus, when Welzer, a German sociologist, writes in *Climate Wars: What People Will be Killed for in the 21st Century* that "violence occurs when there is pressure to take action that will produce results. If these are not forthcoming, new forms of violence are devised – and, if found to be effective, are repeatedly applied", it is evident that the template of The Social Collapse belongs to a repertoire of imaginaries that also exist beyond cli-fi (4).

The judgment: the nonhuman world strikes back

This is also true of the narrative template "The Judgment". However, this template may have a bigger presence in fiction than in non-fiction, since it draws on an animism that modern thought has been eager to overcome. It is thus in contrast to this eagerness that beneath a number of worlds appearing in cli-fi from the Global North, one finds a narrative template driven by the notion of a nonhuman retaliation. According to this template the nonhuman world will at some future point revolt against the devastation and death caused by anthropogenic global warming. At this point the nonhuman world will rise to judge its perpetrators and through disasterous events (e.g. floods, heatwaves, animal attacks on humans) call humanity back to order, that is, back to a state of humility, where an ontology of symbiosis will impose itself through a shift from anthropocentrism to ecocentrism. For example, one can detect such a humility at the end of Roland Emmerich's blockbuster film *The Day After Tomorrow* (2004), when in a televised speech the American president addresses the world's population with the following words: "These past few weeks have left us all with a profound sense of humility in the face of nature's destructive power. For years, we operated under the belief that we could continue consuming our planet's natural resources without consequence. We were wrong. I was wrong" (Emmerich). Before I describe more thoroughly how this monolog fits with the way the film uses the narrative template of The Judgment in a broader sense, let me first, however, turn to a much less known cli-fi: the novel *Gaia Weeps* (1998) by American author Kevin E. Ready. *Gaia Weeps* is by no means a literary masterpiece. In fact it integrates, rather unashamedly, a number of clichés into its plot, which is exactly what makes it a very useful example in this context. In *Gaia Weeps*, a number of minor disasters caused by anthropogenic global warming strike all over the world before a giant tsunami finally covers large parts of the world's populated areas. This event is described as "Mother Nature's, Gaia's revenge for the Industrial Revolution and the myriad of other insults mankind had exposed her planet to" (384). Indeed, the reader is here presented with an animistic intervention that causes major transformations. Thus, on the final pages of the novel, one reads how humanity is being "taught a lesson", while Gaia is restoring "the equilibrium of the planet, cleaning the house after the excesses of her children" (386). In other words: again we end up with humility, as humanity has been wiped clear of its sins in order to arrive at a new order, where it treats the nonhuman world with proper respect. In fact, rather obviously, the deluge in *Gaia Weeps* has the same function as the deluge in a number of early myths like the biblical deluge myth or the myth of Atlantis. However, in contrast to these myths it is not God, who is the final deliverer of a punishment which is also a judgment, but Gaia, Nature, the nonhuman world itself. Similarly, it is no longer the rules of God which have been trespassed, but the rules of Gaia, Nature, the nonhuman world in the sense that an aggregate of nonhuman beings strike back against humanity in order to secure its long term survival.

The cli-fi-thriller *The Swarm* (2004) by German author Frank Schätzing is another cli-fi that draws on this kind of plot. But it is to *The Day After Tomorrow* I will now shortly return. In Emmerich's movie no clear reference to a Gaia-figure is made. This does not mean, however, that the template of The Judgment is not the orchestrator of the plot. In fact, what we are presented with is just another, more complex kind of Gaia-representation than the one we meet in cli-fi such as *Gaia Weeps* and *The Swarm*. In contrast to the unified intentionality, which in these two fiction turns the nonhuman world into a collective war-machine, we encounter a more fragmented kind of war-machine in *The Day After Tomorrow*. This is visible in one of the most suspenseful scenes of the movie, where three of the main characters are chased through the streets of New York by a storm that kills everything, entering its eye. During this chase the soundtrack of the movie is dominated by the snarling of something that sounds like a very large and angry animal. In other words, one encounters here the vindictive nonhuman world in the form of a zoomorphic, monstrous weather phenomenon.

The conspiracy: Manipulation and cover-up

Moving on to the next narrative template, it will hardly come as a surprise to the readers of this anthology that the climate skeptical comprehension of the science and politics of anthropogenic global warming also appear as a plot-structuring element in cli-fi. However, before I give a short account of how the narrative template of "The Conspiracy" is utilized in what can be described as the skeptical climate fiction par excellence: Michael Crichton's thriller *State of Fear* (2004), I would like to draw attention to this thriller's historical context. It is thus interesting that *State of Fear* along with the equally skeptical thriller *Prophezeiung* (2011) by German author Sven Bötcher marks a turn in the history of cli-fi. Until *State of Fear* the narrative template of The Conspiracy is only used to point to the overlooked danger of anthropogenic global warming. Thus, beneath the cover-up in cli-fi thrillers such as Herzog's *Heat* and Rock Brynner's *The Doomsday Report* (1998) one finds the alarming truth that extreme climatic devastation is a real and very serious threat to human existence. So when one encounters climate skeptical plots in *State of Fear* and *Prophezeiung* it can be regarded as a sign of how anthropogenic global warming has gone from being a peripheral scientific phenomenon to a mainstream phenomenon over the course of the last 50 years. In *State of Fear* this historical circumstance is mirrored by the thriller's paranoid tone and characters. Crichton's thriller is in this sense the perfect example of how conspiracy theories, in the words of American cultural theorist Fredric Jameson (1993), tend to appear as "degraded attempt[s] [...] to think the impossible totality of the contemporary world system" (38). Thus, in Crichton's cli-fi thriller all major institutions are repeatedly portrayed as being part of a grandiose cover-up of the fact that future global warming will essentially be harmless. In both the novel and in Crichton's afterword (intended to make clear "where, exactly, the author stands on these issues"!) it is repeatedly suggested that the notion of anthropogenic global warming in reality conceals a coup orchestrated to do away with the freedom embedded in the American way of life (676). However, it is in the writings of Karl Popper and not in those of Jameson that one finds the clearest trace of the historical origin of the template which Crichton's thriller employs. In his essay "Prediction and Prophecy in The Social Sciences" (1948) Popper points out how "the belief in the Homeric gods whose conspiracies were responsible for the vicissitudes of the Trojan War" has not entirely deteriorated (341). Rather, according to Popper, "the place of the gods on Homer's Olympus is now taken by the Learned Elders of Zion, or by the monopolists, or the capitalists, or the imperialists" (342). Put differently, what these remarks lay bare is thus a connection between the ancient idea that God(s) secretly orchestrated the lives of humans and the idea that the scientific community, “the deep state” or other kinds of wicked “puppet masters” use anthropogenic global warming as a manipulative tool. Indeed, Popper’s Greek example can fairly easily be supplemented with a biblical one, as in the Christian tradition, the wicked puppet master is personified in the figure of the Devil and his devilish plots.

The loss of wilderness: The end of nature

I am aware that the narrative templates I have introduced so far may appear familiar or even clichéd. But at the same time this familiarity also underpins my argument that we are here dealing with bricolage in the form of adjustments to narrative templates that have existed since the dawn of Western civilization. This is certainly the case with the template I will turn to now. Beneath a number of worlds appearing in cli-fi from the Global North one thus finds the template of “The Loss of Wilderness”. According to this template anthropogenic global warming will destroy the last wildernesses on Earth. In other words we here encounter a theme that American environmentalist Bill McKibben highlighted as the end of nature as early as 1989. McKibben noted (2003) that “we [humans] by changing the weather make every spot on Earth man-made and artificial” (60). This theme reappears in *The Ice Lovers* (2009) by Canadian author Jean McNeil and *The Lamentations of Zeno* (2016) by German-Bulgarian author Ilija Trojanow. In these two cli-fi we meet main characters, who find it increasingly impossible to live in a world where humans not only “destroy everything aligned with nature” (Trojanow:7), but are more specifically also in the process of creating “a world without ice” (McNeil: 301).

However, as this process is in Antarctica only at an early stage, the main characters in both novels initially experience Antarctica as a sanctuary – or as the continent is described in *The Ice Lovers*: “the largest and most empty wilderness on the planet” (McNeil:145). But this is only to begin with. Gradually their excitement is ruined, as they bitterly experience how their beloved continent, and its animal life is changing due to the impact of anthropogenic global warming and other kinds of human interferences. Whereas the main characters in both novels treat the continent with a deep sense of awe and wonder, the general human mode of being they encounter in Antarctica can – with a term borrowed from Martin Heidegger – best be described as “calculative” (1966:56). This basically means that Antarctica is treated as a giant stockpile of resources “standing in reserve” (Heidegger 1977: 17) – or as described in *The Lamentations of Zeno*: a place, where “people are just biding their time waiting for the day they’re allowed to drill for oil” (105). This scenario is not only integrated into the two novels as a symbolic culmination of a loss of wilderness that is already taking place through the accelerating effects of anthropogenic global warming. The template of The Loss of Wilderness is in both novels also utilized to foreshadow a future in which humanity do not belong. In both novels, the end of wilderness is thus framed as the final, irreversible transgression of tipping points that will drive humanity extinct. In this way both novels symbolically embed the consequences and fate of a collective death-drive within humanity that will in the end cancel itself out. As both novels end with their main character committing suicide their death becomes a synecdoche of the general suicide humanity has initiated. As such the plots of both novels can easily be accused of being enactments of a misanthropic sentiment, which some philosophers have found to be intrinsic to ecological engagement (e.g. Ferry 1995; Bruckner 2013). However, the two novels also display a deep love toward the nonhuman worlds that is bound to mix with elegy in the face of climate devastation and ecocide. In this regard, there is a clear connection between the narrative template and the biblical tale of the eviction from paradise. In both novels the wilderness of Antarctica is thus depicted – through the experiences of the main characters – as a paradisiacal garden capable of restoring human and nonhuman harmony. But as the ice melt, and the search for oil and tourism disturb the continent's ecosystem, the fading and gradual disappearance of this capability become the primary theme of both novel. This means that it is not only the loss of wilderness that the novels lament, but even more so the simultaneous loss of its capacity to fundamentally transform human beings (i.e. reconnect humans with nature and thereby transform their treatment of the nonhuman world). In this regard, misanthropy merely represents one sentiment in a much larger web of affects and understandings, accompanying the dominant templates in the cli-fi utilizing The Loss of Wilderness as their template.

The sphere: Artificial atmospheres

This also comes to light in the fifth narrative template: "The Sphere". According to this template anthropogenic global warming will at some point in the future necessitate the creation of artificial atmospheres. In this regard, the template is linked to the classical sci-fi motif of terraforming, but this link is in itself the product of a connection that according to German philosopher Peter Sloterdijk (2014) goes all the way back to the biblical story of Noah and the birth of "the notion of a group's self-harboring and self-surrounding in the face of an outside world that has become impossible" (237). In any case, it would be reckless to frame the narrative pillars that cli-fi in the Global North have thus far relied on, without giving room to more techno-optimistic narratives than those enabled by the templates presented so far. Even though the template of The Sphere can be found in a broad mix of literary and cinematic cli-fi, this is still very much down to the work of one author. The award-winning American author Kim Stanley Robinson has not only written the arguably most ambitious literary work of cli-fi so far produced in the shape of the trilogy of novels *Science in the Capital* (2004, 2005, and 2007). He has also, in this grand work, as well as in more recent novels such as *2312* (2012), *New York 2140* (2017), and *the Ministry for the Future* (2020), relentlessly depicted worlds wherein geoengineering and atmo-technical design have become the preferred human tools against an increasingly inhospitable climate.

That said, Robinson's novels should not be dismissed as neoliberal fantasies of high-tech futures. In fact, in both *Science in the Capital, 2312, New York 2140* (2017), and *the Ministry for the Future* (2020) the neoliberal project of draining the world for profitable resources before abandoning its ruins in exclusive arcs – think here of the space station Elysium in the film *Elysium* (2013) or the spaceship at the end of the film *Don't Look Up* (2021) – is exactly the project that the major protagonists inexhaustibly fight. In *Science in the Capital* this schism is particularly visible in the depiction of the republican administration, holding the presidency at the beginning of the trilogy, and the democratic administration, holding the presidency at its end. Whereas the plan of the republican administration is framed as a "raiding of the world" that will leave the large majority of humanity in wreckage and the "oil-and-guns crowd" responsible for the wreckage in "bubble fortresses", the plan of the democratic administration is framed completely different (Robinson 2005:472). That is as a world-building, which will be "scientific as can be, but only if you understand science as a way of being together, an ethical system and not just a method for seeing the world" (Robinson 2005:472). In other words, we here find a major conflict over what should be the social design of the atmospheric construction that is to replace the air-conditioning system of the original biosphere: a conflict between, on the one side, exclusive bubbles, and on the other side: an all-encompassing and therefore radically terraformed globe. In *2312* this conflict is still present, although it is in a sense turned upside down. Here, we are told how a large part of humanity has left Earth due to the devastating effects of accelerating anthropogenic global warming and settled in technologically produced spheres on the planets and moons closest to Earth. These spherical communities, depicted as fragile bubbles of inner beauty in an otherwise unsurvivable environment, have to a large extent abandoned the political, social, and economic structures, dominating on Earth. However, as the novel progresses it becomes apparent to these communities that neither the political, social, and economic differences nor their physical detachment are a long lasting guarantee for their well-being and safety. The problem of Earth must be dealt with, as the neoliberal capitalism dominating there still poses a threat for common survival and is furthermore also the cause of extreme devastation, inequality, and loss of biodiversity. So, a "revolution" is initiated through a rewilding of Earth (Robinson 2012: 393). Species long extinct on Earth, but still existent in spheres on other planets due to gene-manipulation are in large numbers literally dropped over Earth in a parachute invasion that eventually leads to basic political, social, and economic changes.

One may laugh at the vividness of these strange events, but what cannot be taken away from Robinson is his engagement in generating new visions for a future that, to follow the thoughts of Italian philosopher Franco "Bifo" Berardi (2011), has, in a way, seized to exist. At a time when the promises given by industrial modernity of "an ever progressing development" have to a large extent reversed into dystopia, Robinson's cli-fi novels at least explore possibilities of other futures ("Bifo" 2011:18). That said, what I wish to imply by drawing this connection between the visions in Robinson's literary works and Bifo's critical analysis of the dire state of the global project of (eco)modernist capitalism is not that we here have a perfect match in the shape of new fantasies ready to replace those that the (eco)modernist project of capitalism has exhausted. Rather, I am drawing this connection in order to showcase how the repertoire that I have delineated does not necessarily protect and conserve the status quo. The fact that authors like Robinson draw on a culturally common repertoire of narrative templates in their world-making does not exclude the possibility that their works can inspire change through imaginative innovations. On the contrary, cultural history here comes to life as the fragments of bricolage that enable the composition of new constellations, new sketches of futures we may or may not decide to appropriate in our search of sustainable futures.

Concluding remarks

To sum up, the argument that cli-fi from the Global North draw on a repertoire of old narratives in its representations of climate changed futures has thus been a key point in my exploration. This is at once helpful and problematic in a world facing the unknowns of the Anthropocene. On the one hand, it is certainly important that human beings are capable of coping with uncertainty and insecurity by way of already well-established cognitive schemes at this difficult time in human history. On the other hand it is also a weakness if we really need, as Latour claimed, to destroy the ideal ideas of the globe that have long been taken as fundamental political and cultural truths. Most prominently, this includes the ideas of endless economic expansion and freedom of consumption, as around the globe these two ideas continue to dominate and feed into political reforms, business models and individual and social modes of living. At a time, when it is becoming increasingly apparent that these ideas are, in the long run, incommensurable with a good future for the human species and many other forms of life, the task becomes – as British philosopher Kate Soper formulates it – to "challenge the supposedly natural evolution of both the capitalist growth economy and the consumer culture it has created, to undermine the sense that this development has been essential to human well-being, and to argue that we will prosper better without it" (13). Here cli-fi has something valuable to offer, as it not only reproduce past ways of understanding the world, but also readjust them to present conditions hence adding new layers to the narratives that drive the world-building of societies and humans forward. This does not mean that we can learn what the future actually has in store for us by reading and watching climate fiction. The futures we encounter here will never be anything more than constructed scenarios. But as these scenarios give birth to a whole catalog of new ways of sensing and understanding different variations of our climate changing world, they also have the power to influence the constructions of new personal and collective truths.

At the beginning of this chapter, I therefore raised the question: How have literary and cinematic fiction from the Global North (i.e. North America, Australia, and Western Europe) engaged with the reality of rapidly accelerating reality of anthropogenic global warming? My answer to this question has taken the form of an introduction to five narrative templates that I have found to dominate in cli-fi from the Global North. I have identified these five templates in a sample of more than hundred (primarily anglophone) literary and cinematic works of cli-fi that have been published during the last five decades. But in the short format of this chapter, I have been confined to inductively demonstrate their existence in a much smaller subset of selected works. And even within this format my reduction of the rich field of cli-fi to five narrative templates undeniably downsizes the complexity and diversity of the selected works and consequently eclipses an array of important differences and details. I have, so to speak, only sought to map the highways and not the sideroads of the field. I would therefore like to end by reminding my readers that the dissemination of scientific findings is always a construction – or "composition" as Latour (2005) would have said – involving both the inclusion and exclusion of knowledge (254). There is simply no scientific dissemination without selection, merely different methodologies. In this sense, my presentation of the five narrative templates should be seen as nothing more than an interpretive grid that seeks to make order out of chaos, that is, provides an analytical overview of an otherwise deeply complex and diverse cultural phenomenon.

References

Andersen, G. (2016). Cli-fi and the uncanny. *ISLE: Interdisciplinary Studies in Literature and Environment, 23*(4), 855–866.

Andersen, G. (2020). *Climate fiction and cultural analysis: a new perspective on life in the anthropocene*. Routledge.

"Bifo" Berardi, F. (2011). *After the future*. In G. Genosko & N. Thoburnn (Eds.), AK Press.

Bracke, A. (2018). *Climate crisis and the 21st-century British novel*. Bloomsbury Academic.

Bruckner, P. (2013). *The fanaticism of the apocalypse: save the earth, punish human beings*. (S. Rendall, Trans.). Polity Press.

Clark, T. (2015) *Ecocriticism on the edge. The anthropocene as a threshold concept*. Bloomsbury.

Crichton, M. (2005). *State of fear*. HarperCollins Publishers.

Emmerich, R. (Director). (2004). *The day after tomorrow*. 20th Century Fox.

Ferry, L. (1995). *The new ecological order*. (C. Volk, Trans.). Chicago University Press.

Gadamer, H. (2006). *Truth and method*. (J. Weinsheimer, Trans.). Continuum.

Geertz, C. (1973). *The interpretation of cultures: selected essays*. Basic Books.

Goodbody, A., & Johns-Putra, A. (Eds). (2018) *Cli-fi: a companion*. Peter Lang.

Heidegger, M. (1966). *Discourse on thinking*. (J.M. Anderson & E.H. Freund, Trans.). Harper & Row.

Heidegger, M. (1977). *The question concerning technology and other essays*. (W. Lovitt, Trans.). Garland Publishing Inc.

Heidegger, M. (2001). *Being and time*. (J. Macquarrie & E. Robinson, Trans.). Blackwell.

Herzog, A. (1989). *Heat*. Tudor.

Jameson, F. (1993). *Postmodernism, or the cultural logic of late capitalism*. Verso.

Johns-Putra, A. (2016). Climate change in literature and literary studies: from cli-fi, climate change theater and ecopoetry to ecocriticism and climate change criticism. *Wiley Interdisciplinary Reviews. Climate Change, 7*(2), 266–282. https://doi.org/10.1002/wcc.385

Johns-Putra. A. (2019a). *Climate change and the contemporary novel*. Cambridge University Press.

Johns-Putra, A (2019b). *Climate and literature*. Cambridge University Press.

Latour, B. (2005). *Reassembling the social*. Oxford University Press.

Latour, B. (2013, March 1). *The anthropocene and the destruction of the image of the globe* [Video]. YouTube. www.youtube.com/watch?v=4-l6FQN4P1c>.

Latour, B. (2014). *On some of the affects of capitalism*. bruno-latour.fr. www.bruno-latour.fr/sites/default/files/136-AFFECTS-OF-K-COPENHAGUE.pdf

Lévi-Strauss, C. (2004). *The savage mind.* (G. Weidenfeld, Trans.). Oxford University Press.
McKibben, B. (2003). *The end of nature*. Bloomsbury.
McNeil, J. (2009). *The ice lovers*. McArthur & Co.
Mehnert, A. (2016). *Climate change fictions. Representations of global warming in American literature.* Palgrave Macmillan.
Popper, K. (2000). *Conjectures and refutations*. Routledge.
Ready, K., E. (1998). *Gaia weeps.* Saint Gaudens Press.
Renfroe, J. (Director). (2013). *The colony* [Film]. RLJ Entertainment.
Robinson, K. S. (2005). *Fifty degrees below*. HarperCollins Publishers.
Robinson, K. S. (2012). *2312*. Orbit.
Sloterdijk, P. (2014). *Globes. Spheres II.* (W. Hoban, Trans.). Semiotext(e).
Soper, K. (2020). *Post-growth living: for an alternative hedonism*. Verso.
Theroux, M. (2009). *Far north*. Faber and Faber.
Trexler, A. (2015). *Anthropocene fictions: the novel in a time of climate change*. Virginia University Press.
Trojanow, I. (2016) *The Lamentations of Zeno*. (P. Boehm, Trans.). Verso.
Welzer, H. (2012). *Climate wars: why people will be killed in the twenty-first century*. (P. Camiller, Trans.). Polity Press.

5 Green populism

Counterpublics and the formation of counterknowledge

Óscar García Agustín and Isabel Jerne

Introduction

In a time when the relationship between scientific explanations and everyday experiences is being questioned, environmental concerns and science have become ever more politicised. Science has played a significant role in strengthening social consensus, social cohesion, and common understanding of phenomena that affect people. However, popular and political debates on controversial topics often involve invoking scientific judgements. Climate change, along with other highly debated issues such as vaccines, abortion, and evolution, is frequently at the centre of such debates. With the rise of populist parties in recent decades, the role of populist and anti-establishment views in climate change debates has become increasingly complex and varies across different contexts (Hultman et al., 2020; Jylhä & Hellmer, 2020). Given that populism and environmentalism are both contemporary and controversial concepts, it is not surprising that increasing scholarly attention has focused on examining environmental politics through the lens of populism and vice versa (Sconfienza, 2022).

In this chapter, we refer to "green populism" as the intersection of populist narratives, themes, and strategies to address climate change and environmental issues. Depending on the context, scholars have defined the phenomenon as "green populism" (Davies, 2020), others as "eco-populism", "environmental populisms" or "populism of environmental movements" (Sconfienza, 2022), acknowledging the variety of contexts and ways in which the phenomenon can develop. In fact, populist and anti-establishment rhetorics have not only been adopted by left-wing and right-wing political parties but have also been adopted by civil society to appeal to a broader audience or to generate dissent.

At the same time, while science is expected to be universal and indisputable, aspects of it have nonetheless been questioned and challenged in the public sphere; a public sphere that is not simply a single realm based on an equal "principle of universal access" (Habermas in Bricker, 2019, p. 682), yet one that is characterised by multiple competing publics; a public sphere that no longer facilitates rational discourse and consensus-building among equals, but a public where counterpublics operate in opposition to dominant discourses and power structures (Fraser, 1991). The multifaceted nature of the public sphere leads to the rise of what Fraser (1990)

DOI: 10.4324/9781003479550-6

defined as "counterpublics" (Warner, 2002), where multiple "counter" civil societies disrupt the traditional Habermasian "monolithic" understanding of a homogenous and deliberate public sphere (Bricker, 2019, p. 682).

In this chapter, we depart from the idea that in the context of climate transformation, the public sphere has become an arena of contestation, where movements, individuals and groups challenge each other and articulate their perspectives and interests either by recognising the legitimacy of others (agonistic way) or by eliminating others (antagonistic way) (Agustín, 2017; Mouffe, 2007). We claim that "green populist" movements and counterpublics are strongly intertwined, reinforcing their identity. Moreover, because of the role of science in climate change, "green populist" movements enhance the formation of "green knowledge" and activate "repertoires of knowledge practices" to produce change. The goal of this chapter is to analyse and conceptualise the interplay between green populism, counterpublics, and their relation to scientific discourses and practices. Our research question is then: *How do green populist groups form counterpublics to articulate their discourse on climate change and their reconstruction of knowledge?* More specifically, the main research question unfolds three sub-questions reflecting three dimensions of green populism:

RQ1: *What are the main counter discourses used by green populist movements to enhance the opposition between the people and the elites? (Discourse)*

RQ2: *How do left-wing and right-wing approaches on environmentalism differ across green populist counterdiscourses? (Ideology)*

RQ3: *How is scientific knowledge addressed to strengthen social consensus amongst the 'pure people'? (Knowledge)*

On one hand, we analyse reactionary or right-wing forms of counterpublics, which tend to oppose climate action or display scepticism on climate change. On the other hand, we analyse progressive or left-wing forms of counterpublics that perform climate activism. While both forms of counterpublics reflect critical standpoints towards climate action, they differ in their understanding of scientific knowledge on climate change. Through a comparative analysis, the study illustrates two extreme cases of (green populist) counterpublics from the Danish context: the national branch of Scientist Rebellion, and Klimarealisme (Climate Realism). Both share commonalities: an anti-establishment discourse, as they challenge mainstream political discourses; the use of scientific knowledge to assert their stances on climate change; the essence of counterpublics, which embeds counter discourses that formulate "oppositional interpretations of their identities, interests and needs" (Fraser, 1990, p. 123). Finally, we wish that the chapter will inspire readers to contemplate the idea of "public sphere", reflecting on the question of "how much room there is in the public sphere for counterpublics and non-mainstream opinions".

By answering the above-mentioned questions, the chapter contributes to the conceptualisation of "green populism" and presents a model for its analysis. In this model, we emphasise the formation of counterpublics and counterknowledge as essential for the understanding of how the knowledge produced by the "elites" is rejected and how knowledge (as counterknowledge) is reshaped. As regards the structure of the chapter, first we define "green populism" and, then, the role of counterpublics and knowledge. We introduce our analytical model on "green populism" and the formation of counterpublics and knowledge, and we use a couple of cases to illustrate how such a model would work if applied more extensively and systematically to other cases.

Defining green populism

In the last decade, an increasing number of studies have explored the connection between populist attitudes and environmental issues (Böhmelt, 2021; Bosworth, 2019; Buzogány & Mohamad-Klotzbach, 2022; Davies, 2020; Jylhä & Hellmer, 2020). In this chapter, we define "green populism" as the intersection of populist discourses, rhetorics, and strategies to address climate change and environmental issues. Both reactionary (right-wing leaning) or progressive (left-wing leaning) manifestations of "green populism" depart from the dichotomy "people vs elites" to express their stances on climate change. For the purpose of presenting a concrete understanding of this to the reader, we present a few cases of what we see as cases that fall under the umbrella of green populism. A first example of reactionary green populism is the right-wing leaning party Sweden Democrats, which has carried an anti-establishment discourse against the "ecomodern hegemony" and environmental advocate-parties such as the Green Party (Hultman et al., 2020). The Danish Democrats stand on similar grounds: echoing Donald Trumps' [CE: change to Trump's] populist rallies, the leader of the Danish Democrats organised a CO_2-rally in April 2024, where 1500 gathered in festivity among tractors and a rodeo bull to oppose a CO_2 tax on agriculture (Würtz, 2024). Within the left-wing spectrum, we argue that the Danish political party Enhedlisten adopts a populist narrative that frames multinational corporations against people and the planet in its advocacy for climate action (Enhedslisten – the Red/Green Alliance, n.d.). This positioning characterises Enhedlisten as an example of progressive green populism. It presents an anti-elite discourse, challenging the influence of major corporations, conservative governments, and the neoliberal policies of the EU – being held responsible for impeding progress in climate mitigation efforts (Enhedlisten – the Red/Green Alliance, 2019)" [CE: delete " here].

Despite the presence of green populism in the right-wing context as well as in the left, existing literature exploring the environmentalism-populism relationship has mainly focused on right-wing ideologies (Hultman et al., 2020; Jylha & Hellmer, 2020; Lockwood, 2018). This literature presented that populists are generally more sceptical towards science and political institutions (seen as the "elites"), and hence tend to deny climate change (Huber et al., 2020). It has also been discussed that populists do not reject environmental debates as such, but instead they contest the environmental policies because these are pursued by technocratic and/or transnational elites (Böhmelt, 2021). Furthermore, while scholars like Bardin and Sigachev (2020), Bosworth (2020) have produced some literature on how left-wing or progressive populism relates to environmental issues (e.g., drawing on Podemos in Spain, and on Five Star Movement in Italy), the literature on progressive green populism is scarcer in comparison to the one right-wing forms (Buzogány & Mohamad-Klotzbach, 2020). The state of the art faces another issue: much of it concentrates on populist party politics, particularly in the US and Western Europe contexts (Böhmelt, 2021). This focus overlooks civil society and broader forms of green populism. While some literature explores the connection between climate policies and communities' responses (Batel & Devine-Wright, 2018), it remains in its infancy and fails to address populist civil society discourses as integral components of green populism practices. This incomplete understanding of green populism runs the risk of overlooking its geographical, spatial, and historical content, picturing the phenomenon as limited to the nation-state while excluding local, "regional and transnational movements" (Bosworth, 2020, p 1), e.g., Extinction Rebellion. Another shortcoming of this literature gap is the inability to theoretically account for the distinction between left forms of populism from those of the right (Bosworth, 2020), as well as for other minor variants such as green "populist counterpublics" (Agustín & Nissen, 2022, p. 745).

Recently, a more critical approach to understanding the connection between populism and environmentalism was given by Sconfienza (2022). Sconfienza (2022) addressed the complexity of the concept, and thus the challenge of capturing all the intersecting junctures between populism and environmentalism. By reviewing four different cases of environmental populism (right-wing populism and climate denial; resource populism; eco-populism; the populism of environmental movements), he demonstrated that environmental populism, which we understand as "green populism", is an intersection of multiple and very diverse junctures (dynamics, actors, framing of environment, framing of people). Because of this complex intersectionality, Sconfienza (2022. p. 231) suggested dropping one singular meaning to the concept of "environmental populism" in favour of "environmental populisms", by instead specifying the diverse junctures that make the phenomenon, movement, party, or event populist.

Given that the state of the art research on green populism is at its early stage of development, yet also acknowledging that green populism combines two concepts that already have vague boundaries (Buzogány & Mohamad-Klotzbach, 2020), our attempt is to contribute to the conceptualisation of green populism. We aim to do so by presenting and analysing how left-wing and extra-parliamentary forms of green populism carry their environmental agendas.

Counterpublics and the formation of green counterknowledge

As different forms of green populism oppose each other (reactionary vs progressive), the debate on climate change has developed a polarisation on environmental stances, affecting the homogeneity of the public sphere. In other words, opposing forms of green populism (such as reactionary and progressive variants) contribute to polarisation on environmental stances – dividing public discourse, making it less unified and inclusive. Since the goal of the chapter is to look at the interplay between green populism, counterpublics, and their relation to scientific discourses and practices, a conceptualisation of counterpublics in relation to scientific knowledge is necessary. We do this by combining the concept of counterpublics with the idea that knowledge production contributes to the creation of new action strategies, the spread of alternative imaginaries and a reform of the status quo (della Porta & Pavan, 2017).

Nancy Fraser (1990) built on Habermas' notion of the public sphere by arguing that there are multiple publics, each with distinct perspectives and interests, introducing the notion of "counterpublics". Fraser described subaltern counterpublics as "parallel discursive arenas where members of subordinated social groups invent and circulate counter discourses to formulate oppositional interpretations of their identities, interests and needs" (p. 123). As elaborated by Agustín and Nissen (2022), these discursive arenas are "subaltern" due to the subordinate nature of the groups that produce the discourses; "counter-"because they embed a function of contestation; and "-publics" because they are part of the public sphere and have a "publicist orientation" (p. 743).

As mentioned, Fraser (1990) challenged the Habermasian notion of a singular, homogeneous public sphere and called for a more nuanced understanding of public discourse that acknowledges the plurality of voices and experiences within society. More recently, Habermas (2022) acknowledged that the public sphere has undergone a transformation because of the growing role of new media in the shaping of public opinions. He reflected that the plurality of public opinions leads then to "*enduring* dissent in the public sphere", increasing the competition between parties, as well as controversies among experts (p. 152). Online media empower all users as authors, opening a space for the easy distribution of low-quality information as well as fake news. Unlike the "old media", the content of online mass media is unregulated, because "professional filters are lacking" (Habermas, 2022, p. 159). For example, platforms like Twitter enable all users to share their opinion and become potentially authors of (fake) news. This leads to a "new" type of opinion formation, one that is sustained by mass media outlining alternative proposals, opposing positions and contributions within the public (Habermas, 2022).

The shift from offline professionally produced news to online and more amateur media, opens space for voices that may share alternative views, which can become influential online (Jackson & Foucault Welles, 2015). In a more positive tone, Castells (2012) explored the role of the Internet media in collective networked citizen debate and decision-making to challenge the messages of dominant social structures. These online counterpublics or "networked counterpublics", indirectly enhanced an egalitarian and unregulated relationship with the recipients, due to the rejection of traditional structures of power (Castells, 2012). For example, tweets can be commented on by readers, enabling a real conversation between the audience and the writers, challenging traditional structures of power.

To summarise, we see that online media has enabled the emergence of multiple counterpublics that claim the expansion of the public sphere or censor their exclusionary function. In the context of climate change, counterpublics openly challenge the "dominant knowledge" inherent to the mainstream public sphere (Agustín &

Nissen, 2022; Jackson & Foucault Welles, 2015). For our purpose, two dimensions are relevant to account for how green populism forms counterpublics: the oppositional identities, articulated in the antagonist form "the people" vs. "the elite" and the oppositional knowledge that legitimates knowledge by questioning the dominant knowledge (which often relies on experts and the scientific community). Therefore, counterpublics enhance the formation of counterknowledge that questions the neutrality of scientific knowledge, as it hides the interests of the political, economic, and media elites. Counterknowledge (Schneider, 2009) implies both questioning the scientific truth and creating one's expertise as one's worldview. In their analysis of repertoires of knowledge practices, della Porta and Pavan (2017) identify three categories: knowledge about oneself, formation and allied knowledge, and political alternatives. Counterpublics promote, in short, a worldview, based on their own (re)production of knowledge that differs from (and is in conflict with) the elite's worldview. In this regard, the intersection between populism, knowledge, and counterpublics becomes clear in relation to the discrepancies and conflicts on climate change.

Model of analysis for green populist counterpublics (ideology, identity, knowledge)

In our conceptualisation of counterpublics as communicative spaces and arenas to shape collective identities and discourses and to produce knowledge, we want to highlight the connection between counterpublics and social movements (Foust et al., 2020). The major focus on online counterpublics does not mean that publics are not hybrid and formed in hybrid arenas (online and onsite). One can be tempted to perceive social media, such as Facebook, Instagram, and Tik Tok, only as reflections of individuals' interaction. While this is true, we want to highlight that counterpublics are shaped in arenas to strengthen collective identities, just like social movements use and diversify their communication in different arenas to expand the scope of their discourse and claims. Thus, social movements and organisations shape their identities, discourses, and practices in both online and offline arenas which are interconnected. Counterpublics are the communicative result of this diversity of interactions and, in our understanding, stress the communicative and discursive dimension of social movements and organisations. Our model of counterpublics consists of three elements: identity, discourse, and knowledge (see Figure 5.1.):

The model above displays counterpublics as defined by ideology, discourse, and knowledge. While ideology refers to the position on climate change, discourse refers to the populist articulation and knowledge refers to the legitimation of the worldview. We maintain a traditional division between left and right (or progressive and conservative) based on the main position adopted in relation to climate change. On the one hand, those who advocate for major and more urgent measures to fight climate change and move towards a green model are associated with a progressive view. On the other, those who diminish the impact of human beings on the climate and deny the immediate consequences of climate change

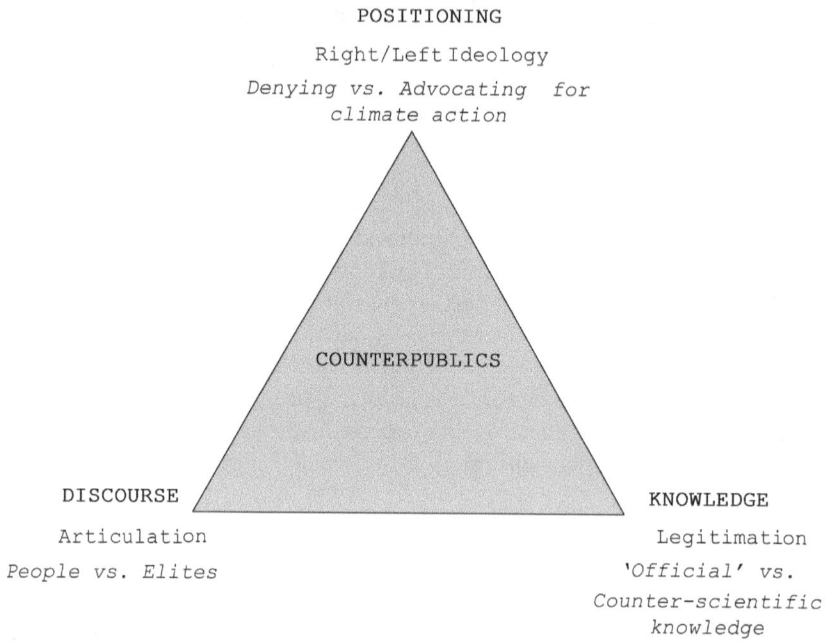

Figure 5.1 The three dimensions of counterpublics.

are placed within a conservative position. Addressing ideology in this way is useful to identify the overall position of social movements and organisations and the role of counterpublics to shape alternative discourses, identities, and views. Furthermore, ideology is important to distinguish between progressive and conservative counterpublics, since the fact that both are populist does not imply that their worldview is similar or identical, rather the opposite. Discourse refers to the populist articulation that provokes an antagonist relation between "the people" and "the elite". This minimal definition of populism (Mudde, 2004) is applied to the position on climate change. The articulation of the divide between the people and the elite can lead to a critique of the elites for not taking the necessary measures or for pushing the green agenda too much. In other words, green populism can become progressive or conservative depending on the ideology (the positioning). What they have in common is, in any case, that the political situation is articulated in terms of a conflict between those who talk in the name of the people and the elites whose agenda aims to reinforce their own interests. Knowledge is connected with ideology and identity, since counterpublics are spaces of production of knowledge to counter the mainstream narrative on climate change (Kaiser, 2017) and "official" scientific knowledge (Hultman et al., 2020). Knowledge is essential to legitimate a worldview and politics that differ diametrically from the scientific knowledge attributed to the elites. Knowledge, therefore, has a dual function: to delegitimise scientific hegemonic knowledge (for lacking neutrality or for supporting the elites' agendas) and to legitimate (as counterknowledge) their own world view which justify other political actions (strengthening climate measures or rejecting them).

Analysing green (populist) counterpublics in Denmark

In this section, we have chosen two cases from the Danish context: the national branch of Scientist Rebellion, and Klimarealisme (Climate Realism). The selection of the cases is due to their different positions about climate change, their populist articulation, and their strong focus on the production of knowledge. It is important to note that both movements challenge mainstream political discourses. Both cases are (green populist) counterpublics from the Danish context, and as such, they share both commonalities: an anti-establishment discourse; the use of scientific knowledge to assert their stances on climate change; the essence of counterpublics, which embeds counter discourses that formulate "oppositional interpretations of their identities, interests and needs" (Fraser, 1991, p. 123). The cases were chosen due to the fact that both movements use the online space as an arena to shape their collective identities and discourses as well as to produce knowledge. As per analysis methods, we relied on hyperlinking, website analysis with an illustrative approach of analysing our data. Rather than presenting a detailed analysis of both groups, the idea is to illustrate how the model connecting counterpublics, populism, and knowledge can be applied to groups that, despite their ideological differences, share a discursive populist articulation. Specifically, for the analysis of the Scientist Rebellion, we heavily relied on their website, X and online links to articles that presented views and opinions shared by the members. As for Klimarealisme, we collected data mainly from their own website and online database, which gathers all their public interventions in diverse media.

Scientist Rebellion

Scientists Rebellion (XR Scientists) is an interesting case for two reasons: it is an international movement, and it is a branch of a larger movement, Extinction Rebellion (XR). Founded in the UK in 2018, XR is an environmental movement, determined to "speak truth to power" and use nonviolent civil disobedience as their main protest strategy. The contentious politics performed by XR in their actions and protests seeks to show "ordinary people that they are the ones with the power, not the elite minority" (Extinction Rebellion, 2020). Climate and ecological emergencies are caused by government, media, and finance. Civil disobedience targets those responsible and aims to have an effective impact on "ordinary people" to know the truth of what is happening with the climate. This conflictual strategy is considered as more efficient than massive parades, since a minority can show the truth to people (Extinction Rebellion, 2020). Consequently, the movement is interested in shaping hybrid counterpublics that connect direct actions with online communication and information both in terms of expanding the effect of the action and of connecting the local actions globally. XR has a progressive vision and claims the need of changing climate politics radically. It produces an antagonism between the elites, not interested in changing those politics, and the people who want to transform society and politics and "save the planet".

The goal of XR activism is to create awareness in "ordinary people" by telling the truth, rather than be a movement composed of "ordinary people". Therefore, the struggle on knowledge becomes one of the most, if not the most, important one. The situation of global emergency has already been pointed out by the scientific community: "Life on Earth is in crisis: scientists agree we have entered a period of abrupt climate breakdown, and we are in the midst of a mass extinction of our own making" (Scientist Rebellion, n.d.). XR Scientists becomes the part of XR that emphasises their identity as experts, scientists, and researchers. The role of scientists is not only to produce knowledge but to take action: "We are a community of scientists from all disciplines working at universities all over Denmark. We take action because we recognise the central role researchers have in spreading awareness and understanding of the climate and ecological crises" (Scientist Rebellion, n.d.). XR Scientists in Denmark is a small group with the intention of raising awareness of the need for urgent climate actions through the use of civil disobedience. For this purpose, they use onsite and online means to give visibility and impact to their message. Despite their conflictual approach against the dominant politics and discourses on climate change, they want to intervene in the public sphere and create consequent political actions based on existing scientific knowledge.

The production of knowledge is, obviously, essential for XR Scientists since it proves the necessity of urgent climate actions. Here, counterpublics are formed to share and expose scientific knowledge that contrasts with the political discourse on climate and to show how the "official" public sphere is excluding voices who question current climate politics. XR Scientists constitutes communicative spaces (a) to make the truth public to all; (b) to open interaction with ordinary people (being the public sphere exclusionary); and (c) to legitimate actions and practices against the interests of the elites. The objective of XR Scientists is to make climate conflicts to transform politics and not to polarise society. Thus, counterpublics target the public sphere by trying to gain influence. The reflection of how and where knowledge is produced, and the critique of the economic and political agenda, are constitutive dimensions of the movement.

James C. Scott (1992) distinguishes between "public" and "hidden transcript". The former reflects the hegemonic vision and excludes any type of disagreement and dissent, while the latter is about ideas, practices, and discourses of resistance that are produced offstage (meaning, without confronting power directly). However, hidden transcripts can come to the surface in the form of conflicts, rebellions, and even revolutions. XR Scientists utilises a similar framework to explain why there is a dissonance between the political-social and the scientific consensus. According to the XR Scientists member, Nikoline Borgermann, we are living in a "collective lie" believing that the goals of The Paris Agreement (limiting global warming to no more than 1.5°C) are achievable. Climate researchers say one thing publicly and this differs from what they say when the cameras are off.

Even United Nations reports do not reflect the scientific conclusions, since they are "corrected by the political elite before being published". XR Scientists has the function of bringing the scientific knowledge (the truth) into the public debate without becoming "public transcript" (i.e., being adapted into the hegemonic discourse). In Borgermann's words: "It is time for the climate scientist to put the facts on the table [lægge svesken på disken] and let the public know how bad it really is" (Borgermann, 2022).

The attempt to bring scientific knowledge into the general public faces the difficulties of questioning the hegemonic discourse (produced and controlled by the elites) and of getting access to the public debates. Concerning the latter, the exclusionary aspect of the climate public debates is stressed and counterpublics prove to be important to find other channels to communicate the scientific truth. Two XR activists talk about their experience in participating in the 2021 "Offshore Summit", a forum for managers from the oil and gas industry invited by members of the Danish Parliament to discuss the role of fossil companies in the green transition. Although the activists paid the fee to participate in the forum, their invitation was withdrawn. They explain their experience as a way of eluding any type of critique:

> However, the evening before the conference we received an e-mail saying that due to 'new circumstances' our registration was no longer valid, and we were not welcome to the summit. We were not alone: activist groups and members in the public sphere [offentligheden], who were known for asking uncomfortable questions to oil and gas industry top executives were also barred from attending. So instead we stood outside DR's Koncerthus -where the event took place- and held a scientific poster session in front of the chain of police officers who protected the building […] We also held this poster session to protest because oil and gas executives -who refuse to listen to science- are given preferential access to politicians, while scientists are denied access.
>
> (Racimo & Portilla, 2021)

This story illustrates the divide between exclusionary publics, formed by politicians and executives, and the counterpublics, formed by scientists/activists. The impossibility of having access to forums, in principle open to the public, reflects the separation between the economic and political action and the scientific knowledge. By making such a gap visible, XR Scientists try to appeal to other publics to create awareness about it.

There is a lack of a common public sphere to deliberate and discuss scientific knowledge, according to XR Scientists. This dissociation between hegemonic publics and counterpublics acquire a strong antagonistic dimension when the movement carries out civil disobedience as a form of protest. Knowledge and action are equally important for XR Scientists, and the content of their social movement consists mainly of diffusion of scientific knowledge and protest actions. One of the major actions in Denmark has been blockading Copenhagen airport to demand banning private jets and to reduce, consequently, the impact of fossil fuels. The police removed the activists and detained some of them. One of the blockaders highlighted the antagonistic dimension of the protest, that XR Scientists defined as "evidence-based":

> I'm here because we are facing a climate crisis. And one of the reasons is that 1% of our population, the ultra rich, are using private jets – they are causing this climate crisis. 1% of our population is causing disasters around the world. We have to stop this. We want to ban private jets.
>
> (XRPathways, 2023)

The categorisation "1%" (as opposed to the 99%), popularised by the Occupy Wall Street Movement, is applied to the ultrarich, the economic elite whose privileges damage climate. Knowledge is the basis for political action and to point out the economic elite as responsible for deepening the climate crisis. The XR Scientists discursive articulation is anti-elitist. As the activists are scientists and experts, rather than opposing the people vs. the elite, they emphasise the negative impact of the economic and political elite on climate. Knowledge is a form of legitimation (the truth) that must be made known to the people. Thus, politicians should adopt the necessary measures to address the urgency of the situation. From excluded positions, XR Scientists want to expand the public sphere and introduce scientific knowledge.

Klimarealisme

Klimarealisme is a small association (267 paying members in 2023) founded by civil engineer Karl Iver Dahl-Madsen. Already in the name chosen for the association, it tries to differentiate itself from "climate scepticism" and "climate denial" by framing its position as "critical". Klimarealisme embodies the remaining opposition to the overall consensus on the ongoing climate crisis (Lystbæk Vestergård, 2022). The existence of the climate crisis is not denied and the idea that it is "partly" provoked by human actions is acknowledged. Thus, the objective of the association is to contribute to a *fact-based* debate on climate and climate politics, as the emergency of a climate crisis is denied, and climate actions are strongly questioned. Despite claiming being party-politically independent, the association has contact with the radical right-wing party Nye Borgerlige (New Right). When talking about its vision, Klimarealisme refers to Geert Wilders' victory in the Dutch elections in 2023 as a proof of the unstoppable advance of their standpoint on climate that is framed in terms of "popular uprising" [folkeligt oprør]: "They [the Social Democrats] have recognized that there will be a popular uprising when the population realises what the green foolishness [tosserier] really costs and for no use whatsoever" (Dahl-Madsen, 2023). In this regard, the association allies with the far-right position on climate and expects that "the people" will be on their side when they realise the inconsistency of climate politics.

The lack of presence of Klimarrealisme in the public debate and the lack of access to social media explain the formation of counterpublics, especially online. The website is central to archive all the information produced by the association, and social media (in particular, Facebook) are used to distribute the content. Knowledge plays a fundamental role to delegitimise scientific knowledge sustaining climate politics (characterised as "belief") and to present their own (counter)knowledge as the basis for a new politics (with e.g., less focus on climate measures and more investment in nuclear energy). Mass media becomes one of the major adversaries because they exclude them from the public sphere and disseminate unnecessary alarmism, according to the association. Klimarealisme proposes their own production of counterknowledge (critical knowledge against the dominant one), they reject the exclusionary role of the public sphere (dominated by mass media), and they denounce the negative consequences of "scientific" knowledge when becoming political action. We see that the production and counter-production of knowledge generates what Habermas (2022) referred to as "*enduring* dissent in the public sphere", increasing the competition between parties, as well as controversies among experts (p. 152)

As mentioned, Klimarealisme does not deny climate change, but this does not make their position less critical against climate politics. The association is aware of the fact that their opinions on climate are considered "conspiracy theories" by some. Therefore, there is an effort to highlight the scientific profile of the group and the expertise of its members. The notion of "realism", included in the name, creates a separation between what is real and what is fictional. "Realism" entails being critical with theories that are not reflecting what is really happening: "Klimarealisme has a research and knowledge-based approach to the climate challenge. This means that we must deal with reality, regardless of whether we like it or not. And being equally critical of all kinds of research results regardless of their direction" (Dahl-Madsen & Pedersen, 2019). Consequently, they oppose "the dramatic and alarmist media-created myths about the climate" (Dahl-Madsen & Pedersen, 2019). Realism is, in other words, the opposite of alarmism; and knowledge is the opposite of myth.

Information and dissemination of knowledge is related to the public sphere and the possibilities of deliberating and presenting (counter)arguments. Klimarealisme criticises the exclusionary function of the public sphere, controlled by mass media. On its website, Jørgen Keincke (2023) states that the climate elite (mass media and politicians) is out of touch with the people, drawing on research showing that Danish people would not pay the price for CO_2 reduction and that they have other priorities. While the media and politicians are victims of climate hysteria, "Danish people are by far more reasonable" (Keincke, 2023). This discrepancy between public and people's opinion is the underlying background for an incisive critique of mass media as the responsible actor for creating alarmism among people and reproducing the views of the elite. The attacks are directed against mainstream media and, in particular, against the Danish public broadcasting corporation, DR. Here an example:

> DR thus contributes in a violent, unequivocal propaganda manner, unchallenged and unambiguously, to promote the dominant climate alarmism to an extreme degree. Children are scared, teachers in schools become cowards and society is cringing under all the climate doomsday talk. People stop watching TV news

> and close their ears to news because of the persistent doomsday tones from the 'elite' regarding climate.
>
> (Hansen, 2023)

The climate "alarmism" is attributed to mass media that promote climate anxiety and unnecessary concerns. A divide between the elite (the promoters of climate alarmism: here, mass media) and the people (refusing to listen to media's propaganda) is presented to emphasise that the "real" information can only be found through other alternative publics, i.e., Klimarealisme, willing to contradict the dominant discourse of the elites. The knowledge produced in the public sphere by mass media is not scientific but hyperbolic since it tries to condition people's behaviour. However, Klimarealisme does not believe that the media achieve this successfully, as proved by people not following the "hysterical" attitudes that the elites try to provoke.

The convergence of alarmism promoted by the media and the highly politicised scientific community is dangerous. The reason is that climate action reflecting this convergence can be harmful for the people. Besides the discussion about the objectivity of scientific knowledge, it is clear that Klimarealisme places the focus on criticising the media's role and political actions. The objective of the association is, consequently, to intervene in the public debate by rejecting climate political actions. This position implies the delegitimisation of scientific knowledge as the basis for ongoing climate politics:

> It is fantastic that the scientific environment around the climate has moved so far that you can take this kind of thing seriously. But it is also dangerous because the green transition, if it goes too far, can cause irreparable damage to our energy supply and thus to society, and ultimately to the individual person.
>
> (Hansen, 2021)

The hegemonic scientific knowledge is blamed for pushing the climate agenda and imposing measures to promote political actions that are going to damage the society and individuals. The consequences of scientific knowledge are opposed to the interests of the people. Therefore, another kind of knowledge needs to be produced to address the climate question more objectively and support the interests of the people.

To Klimarrealisme both mass media and scientific knowledge are the main adversaries for creating an "alarmist" environment and an exaggerated focus on climate political actions. While the media are responsible for strengthening the hegemonic discourse on climate change in the public sphere, scientific knowledge constructs a sense of legitimation of those actions. Since they feel excluded from the public sphere (particularly, by being accused of promoting conspiracy theories), Klimarrealisme contributes to form and develop climate sceptic counterpublics where scientific knowledge and media are openly questioned, and alternative knowledge and online platforms are deployed. Despite not having a strong focus on "ordinary people", the association articulates a significant anti-elitism and frames itself as defender of the individual, in particular, and of the society, in general, i.e., those who are going to suffer from the consequences of climate politics.

Conclusion

The increasing contestation on climate change has led to greater polarisation and the emergence of echo chambers, fostered by use of digital media, to amplify extreme ideas (Barberá, 2020). We have decided to address this phenomenon from the perspective of "green populism", i.e., groups that address the climate issue through the discursive antagonism between the people and the elite. While nature and climate are articulated together with the interests, lifestyle and emotions of the people, the actions of the elites are presented as a threat to the people and nature. Digital media has contributed to the formation of counterpublics where populist discourses and identities on climate can be developed and shape their antagonist position against mainstream or hegemonic climate politics and discourses. This is in line with Habermas' more recent understanding that the public sphere has undergone a transformation because of the growing role of new media. In the online public space, "enduring dissent" has increased the competition between parties, as well as controversies among experts (Habemas, 2022, p.152). In fact, in the formation of green populism, knowledge is a dimension that plays an important role and sometimes it has not been seriously considered (especially when associated with post-truth and conspiracy theories). Whether the climate crisis is acknowledged or denied, knowledge is, in any case, at stake, since the neutrality of scientific knowledge is questioned. Counterpublics enable the questioning of hegemonic knowledge, as well as the political and communicative practices associated with it, and the production of counterknowledge that, besides questioning the scientific truth, develops its own expertise and worldview.

When connecting all these aspects, we propose a model to conceptualise and analyse how counterpublics can be seen as the site of production of identity (progressive or advocate of climate politics and denial or sceptical), of discourse (by articulating an antagonistic relation between the people and the elite), and of knowledge (by delegitimising the hegemonic knowledge and developing knowledge practices that legitimate their own worldview). Therefore, we have proposed a model that includes these three dimensions (ideology, identity, and knowledge) to highlight the importance of ideological positioning in relation to environmentalism, identity in terms of populism, and the formation of knowledge. The focus on ideology as a general category allows us to analyse both counterpublics that advocate for climate action and those that deny its necessity. It also enables an analysis of how the populist discourse articulation and knowledge differs completely in terms of who the elite is and what legitimate knowledge is.

To illustrate the application and possibilities of the model, we have chosen two opposite cases from the Danish context: Scientist Rebellion and Klimarealisme (see Table 5.1). Both share a strong focus on knowledge as the defining element of their identity and actions. Scientist Rebellion is part of Extinction Rebellion but with a strong focus on scientific knowledge by experts and academics. As a

Table 5.1 Scientist Rebellion and Klimarealisme as counterpublics

	Ideology	*Identity*	*Knowledge*
Scientist Rebellion	Advocating for climate change; urgent and more determined political action	Strong academic and scientific profile; critique of the elite (politicians, economic interests)	Bringing the existing scientific knowledge into the public debate; urgent climate action based on knowledge
Klimarealisme	Not denying climate change; denying the necessity of adopting political action	Emphasis on expertise; critique of the elite (mass media, scientific community)	Promoting knowledge that proves that climate action is neither a priority nor a necessity; criticising scientific knowledge and its diffusion by mass media

result, the populist focus on the people, more present in the Extinction Rebellion, is not highlighted very often but instead replaced with the importance of "telling the truth". Anti-elitism is, however, important, and elites are accused of hindering political action because it contradicts the interests of the elite (the few against the interest of the many). Klimarealisme appeals to its own expertise to give voice and visibility to national and international experts that disagree with the scientific knowledge on climate change and the need for immediate political action. The members represent the experts rather than the people, but they expect to gain the support of "ordinary people" when they get access to their knowledge. The anti-elitist element is very accentuated, as the public opinion and the production of scientific knowledge is controlled by an irresponsible elite that is leading the people towards an economic disaster. In consequence, counterknowledge shows that real politics should address people's concerns and needs, and climate is not the one issue that a political agenda should prioritise.

We believe that the analysis of green populism with focus on knowledge and counterpublics can contribute to strengthen the field of studies on the interconnection between climate politics, communication, and the production of knowledge. We have introduced the ideas of "green populism" and "green counterpublics" to stress how populism (as discursive articulation) and counterpublics (as communicative spaces) enable knowledge practices and the formation of their own expertise and worldview. There are other aspects that could be interesting to explore in the future to have a more detailed understanding of this phenomenon, although it has not been part of this chapter: (a) the relationship between populist counterpublics and conspiracy theories and post-truth; (b) the convergence of climate concerns with other topics in the formation of counterpublics (e.g., pandemic or the defence of peace against war); and (c) the interplay between online and physical spheres and their increasing hybridity (from online groups and forums to protests and actions in the street). By expanding towards these topics and counting on more empirical and comparative analyses, we will have a better understanding of how climate politics and knowledge is being contested and what the potentials and risks arising from the different forms of contestation are.

References

Agustín, Ó.G. (2017). European counterpublics? DiEM25, plan B and the agonistic European public sphere. *Journal of Civil Society*, *13*(3), 323–336. https://doi.org/10.1080/17448689.2017.1360233

Agustín, Ó.G., & Nissen, A. (2022). The anti-restrictions movement and the populist counterpublics in Denmark. *PArtecipazione e COnflitto*, *15*(3), 741–760. https://doi.org/10.1285/i20356609v15i3p741

Barberá, P. (2020). Social media, echo chambers, and political polarization. Social media and democracy. In N. Persily & J. Tucker (Eds.), *Social media and democracy: the state of the field* (pp. 34–55). Cambridge: Cambridge University Press.

Bardin A., Sigachev M. (2020). The green discourse as a type of new left populism. *World Economy and International Relations*, *64*(11), 96–105. https://doi.org/10.20542/0131-2227-2020-64-11-96-105

Batel, & Devine-Wright, P. (2018). Populism, identities and responses to energy infrastructures at different scales in the United Kingdom: a post-Brexit reflection. *Energy Research & Social Science, 43*, 41–47. https://doi.org/10.1016/j.erss.2018.05.011

Böhmelt, T. (2021). Populism and environmental performance. *Global Environmental Politics*, *21*(3), 97–123. doi: https://doi.org/10.1162/glep_a_00606

Borgermann, N. (2022). Medlem af Scientist Rebellion: Bag lukkede døre taler klimaforskerne om, at vi lever på en kollektiv løgn. *Klimamonitor*. https://klimamonitor.dk/debat/art9041902/Bag-lukkede-d%C3%B8re-taler-klimaforskerne-om-at-vi-lever-p%C3%A5-en-kollektiv-l%C3%B8gn

Bosworth, K. (2020). The people's climate march: environmental populism as political genre. *Political Geography*, *83*, 102281. https://doi.org/10.1016/j.polgeo.2020.102281

Bricker, B.J. (2019). Scientific counterpublics: in defense of the environmental scientist as public intellectual. *Topoi 38*(4):681–692.

Buzogány, A., Mohamad-Klotzbach, C. (2022). Environmental populism. In M. Oswald (Ed.), *The palgrave handbook of populism* (pp 321–340). Palgrave Macmillan, Cham. https://doi.org/10.1007/978-3-030-80803-7_19

Castells, M. (2012). Networks of outrage and hope: social movements in the internet age. John Wiley & Sons.

Dahl-Madsen, K.I. (2023). Godt nytår til alle klimarealister. *Klimarealisme.dk*, 31st of December. https://klimarealisme.dk/2023/12/31/godt-nytaar-til-alle-klimarealister/

Dahl-Madsen, K.I. & Pedersen, J.O.P. (2019). Det, vi godt kan lide at høre. *Klimarealisme.dk*. https://klimarealisme.dk/2019/09/25/det-vi-godt-kan-lide-at-hoere/

Davies, W. (2020). Green Populism?: action and mortality in the anthropocene. *Environmental Values*, *29*(6), 647–668. Retrieved from https://research.gold.ac.uk/id/eprint/27259/

della Porta, D., Pavan, E. (2017). Repertoires of knowledge practices: social movements in times of crisis. *Qualitative Research in Organizations and Management: An International Journal*, *12*(4), 297–314.://doi.org/10.1108/QROM-01-2017-1483

Enhedslisten – The Red/Green Alliance. (n.d.). Enhedslisten.EU. https://enhedslisten.eu/kontakt/enhedslisten-the-red-green-alliance/

Enhedslisten – the Red/Green Alliance. (2019). *Europæisk klimaudspil* [PDF]. https://enhedslisten.dk/wp-content/uploads/2019/04/A5-folder_Europ%C3%A6isk-klimaudspil_Web.pdf

Extinction Rebellion (2020). Extinction Rebellion blocks News Corps Printworks and demands they 'Free the Truth'. *Extinction Rebellion*. 4th of September. https://extinctionrebellion.uk/2020/09/04/breaking-extinction-rebellion-blocks-news-corps-printworks-and-demands-they-free-the-truth/

Foust, C.R., Pason, A., & Zittlow Rogness, K. (2020). What democracy looks like: the rhetoric of social movements and counterpublics. *Communication Research Trends*, *39*(2), 38–39. www.proquest.com/docview/2430113885?OpenUrlRefId=info:xri/sid:primo&accountid=8144

Fraser, N. (1990). Rethinking the public sphere: a contribution to the critique of actually existing democracy. *Social Text*, *25/26*, 56–80. https://doi.org/10.2307/466240

Habermas, J. (2022). Reflections and hypotheses on a further structural transformation of the political public sphere. Theory, Culture & Society, *39*(4), 145–171. https://doi.org/10.1177/02632764221112341

Hansen, S. (2021). Pseudoforskning. July 7. *Klimarrealisme*. https://klimarealisme.dk/2021/07/07/pseudoforskning/

Hansen, S. (2023). Klima – nødsituation, i DR. Klimarealisme, August 12. *Klimarrealisme*. https://klimarealisme.dk/2023/08/12/klima-noedsituation-i-dr/

Huber R.A., Fesenfeld, L., & Bernauer, T. (2020). Political populism, responsiveness, and public support for climate mitigation. *Climate Policy*, *20*(3), 373–386 https://doi.org/10.1080/14693062.2020.1736490

Hultman, B.A., & Viinikka, T. (2020). The far right and climate change denial: denouncing environmental challenges via anti-establishment rhetoric, marketing of doubts, industrial/breadwinner masculinities enactments and ethno-nationalism. In B. Forchtner (Ed.), *The far right and the environment* (pp. 121–135). Routledge. https://doi.org/10.4324/9781351104043-8

Jackson, S.J. & Foucault Welles, B. (2015). Hijacking #myNYPD: Social media dissent and networked counterpublics. *Journal of Communication*, *65*(5), 932–952, doi:10.1111/jcom.12185

Jylhä, & Hellmer, K. (2020). Right-wing populism and climate change denial: the roles of exclusionary and anti-egalitarian preferences, conservative ideology, and antiestablishment attitudes. *Analyses of Social Issues and Public Policy*, *20*(1), 315–335. https://doi.org/10.1111/asap.12203

Kaiser, J. (2017). Public spheres of skepticism: climate skeptics' online comments in the German networked public sphere. *International Journal of Communication*, *11*, 1661–1682. www.researchgate.net/publication/316170941_Public_Spheres_of_Skepticism_Climate_Skeptics'_Online_Comments_in_the_German_Networked_Public_Sphere

Keinicke, J. (2023). Klimaeliten ude af trit. *Klimarealisme.dk*, 20th of July. https://klimarealisme.dk/2023/07/20/klimaeliten-ude-af-trit/?utm_source=mailpoet&utm_medium=email&utm_campaign=nyhedsbrev-fra-klimarealisme-newsletterposttitle-4

Lockwood, M. (2018) Right-wing populism and the climate change agenda: exploring the linkages. *Environmental Politics*, *27*(4), 712–732. DOI: 10.1080/09644016.2018.1458411

Lystbæk Vestergård, G. (2022). Den sidste klimabenægter. Weekendavisen, 29th of September. www.weekendavisen.dk/2022-39/ideer/den-sidste-klimabenaegter

Mouffe, C. (2007). Artistic activism and agonistic spaces. *Art & Research, 1*. www.artandresearch.org.uk/v1n2/mouffe.htm

Mudde, C. (2004). The populist zeitgeist. *Government and Opposition, 39*(4), 542–563. DOI:10.1111/j.1477-7053.2004.00135.x

Racimo, F. & de Lapuente P., Lopez (2021). Scientist Rebellion: Forskere gør oprør. *Solidaritet*, 15th of December. https://solidaritet.dk/scientist-rebellion-forskere-goer-oproer/

Schneider, U.J. (2009). Counter-knowledge. an update on foucault in the age of mobile communication. In: J.K. Nyíri (Ed.), *Engagement and exposure. mobile communication and the ethics of social networking* (pp. 103–108). Passagen Verlag: Wien 2009.

Scientist Rebellion (n.d.) XR Scientists. *Extinction Rebellion Danmark*. www.xrdk.org/en/science/

Sconfienza, U. M. (2022). What does "environmental populism" mean? mapping the existing debate and a plea for dropping this label. *Populism, 5*(2), 209–235. https://doi.org/10.1163/25888072-bja10035

Scott, James C. (1992). *Domination and the arts of resistance: hidden transcripts*. New Haven: Yale University Press.

Warner, M. (2002). *Publics and counterpublics*. Zone Books. https://doi.org/10.2307/j.ctv1qgnqj8

Würtz, J. (2024, April 11). Inger Støjbergs rally er populistisk – og et "klart sundhedstegn" for det danske demokrati? *DR.DK*. www.dr.dk/nyheder/politik/med-rodeotyr-oel-og-airfryer-giver-inger-stoejberg-sin-politiske-tale-noget-andre

XRPathways (2023). Scientists Rebellion in Denmark blockade airport. *XRPathways*, 10th of August. www.xrpathways.org/scientists_rebellion_in_denmark_blockade_airport

Part 2

Contested science

Conspiracy and counter-knowledge

6 Fighting (for) truth?

Alex Jones, the WHO and the legitimation of conspiracy discourse

Massimiliano Demata

6.1 Introduction

The origins of COVID-19, its short- and long-term effects and its treatment, were (and to a great extent still are) highly contested issues, both on a medical and a political level. Not surprisingly, ever since the beginning of the pandemic in early 2020, all aspects related to the virus have been affected by conspiracy theories, disinformation and false beliefs of all kinds: it has even been alleged that the virus originated from covert operations orchestrated by the United Nations and the World Health Organization (WHO), and that various concealed actors either engineered or intentionally disseminated it.

While they have often been chastised as "bad science" (Harambam & Aupers, 2015) or deviations from accepted norms of rationality or even morality, conspiracy theories have a significant impact in many countries, affecting people's health while also influencing political process. Conspiracy theories are branded as one of the main contributors to, and symptoms of, the so-called post-truth era, a period in which emotions take precedence over facts and empirical evidence in shaping public understanding of social reality, significantly influencing how large segments of society acquire and interpret information (McIntyre, 2018). In the USA, in particular, these theories have managed to reach beyond those fringe groups who usually support them and to gain mainstream status, mainly through former (and newly elected) US President Donald J. Trump, who often promoted conspiracies as part of his own discourse.

Of the many media pundits to peddle conspiracy theories, Alex Jones is certainly one of the most controversial. For decades, Jones has been an active proponent of conspiracy theories. However, during the pandemic, his activities on his platform, InfoWars, were particularly prominent in disseminating disinformation and challenging official medical reports and health policies, especially those issued by the WHO. Additionally, he positioned himself as a purveyor of the "truth", countering the narratives presented by mainstream media and political authorities.

This chapter seeks to address fundamental questions regarding the construction and widespread appeal of conspiracy theories. Specifically, it examines the factors that contribute to their persuasiveness and success, the discursive strategies employed by their proponents to enhance their credibility, and the

DOI: 10.4324/9781003479550-8

mechanisms through which they are effectively disguised to resemble truth. Furthermore, it critically investigates whether conspiracy theories can be definitively situated within the contemporary framework of post-truth discourse. In order to answer these questions, the term *epistemic masquerade* is proposed here to define the way conspiracies are enacted in discourse, whereby they are disguised as truth and legitimized by certain masking and manipulating devices at both the visual and linguistic levels. These devices are shown to be at the root of the communication strategies and discursive repertoire of Alex Jones, perhaps the most influential conspiracy theorist in the USA. I contend that Jones's discourse is paradigmatic of the epistemic masquerade that underpins the nature of conspiracy theories. This chapter examines two videos in which Jones denounces the WHO, portraying it as a component of a broader global conspiracy allegedly orchestrated by the United Nations and other actors. Grounded in a network of conspiracy theories concerning the origins of the virus, the purported dangers of vaccines and an overarching objective to impose global tyranny, Jones strategically constructs his discourse to enhance the plausibility of these claims. Through this rhetorical approach, he seeks to legitimize these theories while securing the trust of his audience.

In the first section of the chapter, I discuss conspiracy theories as discourse. This is in line with my previous work on conspiracy theory (Demata, Zorzi, & Zottola, 2022; Demata, 2024), whereby I consider such theories to be fully fledged, self-contained discourses, claiming the status of (socially constructed) truth in opposition to the official narratives of the media and political establishment. The next sections are devoted to the role of the WHO during the pandemic and to an account of Jones's status as a controversial anchorman who has risen to fame over the years by peddling conspiracy theories of all and every kind. The following section on methodology examines the two interrelated discursive strategies, manipulation and legitimation, that underpin the analysis of the two videos. In the subsequent sections, the analysis and discussion of Jones's videos demonstrate that conspiracy discourse operates as a hybrid, wherein ostensibly scientific sources, alongside manipulated or selectively transcribed official "reports", function as legitimation mechanisms for a series of unfounded claims. The discursive architecture of conspiracy discourse employs a number of persuasive strategies that revolve around the authority of what are presented as scientific sources and evidence. This legitimation and *masquerading* of truth occur because Jones, like all other conspiracy theorists, validates his claims by accumulating evidence through reference to other hyperpartisan (yet deemed to be authoritative) platforms and websites, thus creating an echo chamber in which his views gain traction and are granted "truth" value.

6.2 Truth as (and in) discourse: Conspiracy theories as *epistemic masquerade*

According to a widely accepted definition (e.g. Sunstein & Vermeule, 2009; Douglas et al., 2019; Uscinski, 2020), conspiracy theories constitute interpretative and communicative frameworks that offer alternative explanations to the official accounts of events. For conspiracy theorists, everything stems from an intentional cause and has somehow been controlled and manipulated by obscure and powerful evil actors or groups acting against the people and the common good. The theorists come up with explanations for seemingly unrelated events and see connections where there are none. In this sense, they purport to explain what, in their opinion, can be proved empirically through observable and factual evidence which is instead kept well hidden from people's view. Those who propose or support conspiracy theories deny the validity of the narratives proposed by the "epistemological authorities" (Uscinski, 2020). They therefore counter the supposedly unbiased fact-based claims of social institutions such as the government, scientists, academia and the media. Every aspect of social reality has its own array of conspiracy theories. From politics to science, from entertainment to sport, no field is immune from groups of people theorizing over covert actions performed by hidden actors lurking beneath the surface of official facts or explanations. While they have always existed, conspiracy theories are particularly successful in a context of "epistemological insecurity" (Harambam & Aupers, 2015) or "epistemic instability" (Harambam, 2020), in which the authority of science is undermined and alternative views of reality progressively gain ground. Periods of social, political and economic unrest lead to insecurity and the advent of some unexpected or uncontrollable event serves to undermine social or political certainties.

While it would be easy to dismiss conspiracy theories as false, they in fact provide their believers with a worldview, a set of rules of social behaviour and their own "truth". The current popularity of conspiracies raises important questions as to *why* their discourse is so credible, and how it is structured vis a vis "official" or legitimate ideologies and discourses. The term *epistemic masquerade* is proposed here to define conspiracy discourse and its claim to truth, as the discursive strategies to which conspiracies resort serve as a prop or façade to support the "truth" being promulgated. Conspiracy theories can be understood as challenging what is generally perceived by the majority as established facts or fundamental empirical evidence, including that which is supported by scientific inquiry. However, I argue that these theories exhibit their own internal discursive coherence and logic, which function to construct truth through self-legitimizing discursive strategies manifested within texts.

Epistemic masquerade manifests through three interrelated dimensions, each reflected in the discursive and linguistic structures of conspiracy texts. The first is their *oppositional epistemology*, a fundamental aspect of the ideological framework underpinning conspiracy theories, whereby such narratives position themselves in direct opposition to the "official" or hegemonic ideology. This oppositional stance frequently translates into the political sphere through the second dimension, the *polarized logics of populism*, which channel the epistemic masquerade of conspiracy discourse into social and political praxis. This process unfolds primarily within the third dimension, *a fragmented media landscape*, which provides an optimal environment for conspiracy actors to obscure false claims as truth and cultivate an intensely polarized communicative and discursive space. Indeed, as shown by Mäkinen (2025) in this volume, polarization in communication is a very common feature in the discourses surrounding the COVID pandemic, as, at the time, highly conflictual views from different communities of people became the basis of public debates about the pandemic. These three dimensions are not new in the wide literature on conspiracy theories but I am using them here as the theoretical foundation of the notion of epistemic masquerade in terms of *counter-discourse*, which proposes the truths of social realities in opposition to those supported by official epistemological authorities. This occurs in texts mainly through discursive strategies of *manipulation and legitimation*. These two overlapping strategies mimic those used by the legitimate discourses of the epistemic authorities and create a "discursive illusion" of truth (Bhatia, 2015) which is typical of public discourses and, even more so, of conspiracy theories.

Conspiracy discourse generates an alternative epistemology of social reality. In this regard, it functions analogously to ideology by articulating a system of values and beliefs that diverges from the dominant ideological framework. By doing so, it constructs a worldview positioned in opposition to the allegedly deceptive narratives endorsed and perpetuated by the mainstream (Fenster, 2008; Fuchs, 2021). Conspiracies seem to act, in the Gramscian sense, in support of anti-hegemonic values, in that they oppose the social, cultural and political hegemony of the ruling classes (the "elite", the "establishment" or, in typical conspiratorial jargon, the "mainstream") and its instruments of legitimation, i.e. the epistemological authorities. They do so in the name of the general will, which is discursively constructed as an "us" or "people" allegedly marginalized and violated by those socially hegemonic elements which conspire against it. As argued by Harambam (2020), conspiracy theorists are in direct competition with established science in the battlefield of knowledge contestation, each vying for epistemic authority. In this sense, official science and conspiracy theories are mutually exclusive: they delegitimize each other and share no point of convergence when it comes to their explanation and evaluation of reality. They effectively construct two separate ideologies, each claiming legitimacy and each claiming to be the voice of truth.

The oppositional epistemology proposed by conspiracy counter-discourse is the expression of, and feeds into, an alternative "political epistemology" (Chapelan, 2021: 282), which becomes particularly evident during periods of crisis. This counter-discourse both expresses and feeds into forms of radical far-right populism. In much the same way as populists, conspiracy theorists cater for the sense of disenfranchisement felt by many social groups as they claim oppressed status and pose as the oppressed Other. They do so by attacking any voice of authority and proposing the(ir) truth in direct opposition to that of the epistemological authorities. Indeed, conspiracy theories are both the cause and the effect of what has been seen as an epistemic crisis of democracy, i.e. a growing distrust of any form of widely shared and *agreed upon* knowledge regarding the basic facts pertaining to social reality. Populist and far-right parties and leaders often exploit conspiracy theories to pursue their anti-systemic politics. In fact, populism and conspiracy theories intersect in their fundamental critique of the establishment, which they portray as engaging in clandestine actions that undermine the interests of the people and the collective will. The epistemological authorities which are the target of conspiracy theories are one facet of the establishment which populists rail against as a staple of their trademark politics. The good vs evil polarization embodied in the people vs conspirators opposition typical of conspiracy discourse intersects with that of the people vs elite typical of populism, a moral polarization in and of itself (Bergmann, 2018). Furthermore, the conspiracy claim to truth corresponds to the populist claim to authenticity. Both populists and conspiracy theorists (the two groups indeed often overlap) claim to speak the truth and to represent the pleas and needs of the marginalized and humiliated majority, or "people" (Enli, 2024).

Finally, the success of modern conspiracy theories would appear to be, at least in part, a product of the new media system. Knowledge is now made available through a highly fragmented media system, with mass media and social media competing for audience attention but ultimately intermingling into new forms of media hybridity (Chadwick, 2017). In the digital environment, traditional hierarchies are undermined, and social actors, hitherto voiceless, can now find an outlet for their ideas with a potentially unlimited audience, claiming legitimacy and in turn delegitimizing the all-powerful political and media elite. The result is an "epistemic cacophony, where even the descriptions of basic social realities are often contested" (Dahlgren, 2018: 25) and where "competing versions of knowledge and facts arise, thereby generating incompatible views of reality, cementing and isolating incompatible discursive bubbles, and eroding the grounds for political discussion" (Dahlgren, 2018: 22). Sites and platforms supporting conspiracy theories and disinformation thrive in an environment in which there is no longer a single, idealized and unified public sphere, but rather a multiplicity of "micro" public spheres, each in direct (political and commercial) competition with traditional mass media. In fact, it has rightly been argued that it makes increasingly little sense to speak of a single public sphere as a homogenous entity, as we should rather speak of fragmented or micro-audiences, "in a fragmented field of media content production and circulation" (Blommaert, 2020: 393). This set of alternative, fragmented micro-audiences make up what has been called the "anti-public sphere", i.e. a space (often online) which routinely, by its own nature, opposes the standard values of norms and conventions which are traditionally associated with democratic discourse (Davis, 2021).

6.3 The discursive strategies of conspiracy: Manipulation and legitimation

The three interrelated dimensions of conspiracy as epistemic masquerade described in the previous section find concrete realization in the language used in conspiracy discourse. As discussed in the author's previous work (Demata, Zorzi, & Zottola, 2022; Demata, 2024), in order to claim legitimacy, conspiracy theories construct a discourse which stands as an alternative to the legitimate one. Discourse is here intended as a set of "context-dependent semiotic practices" which are "socially constituted and socially constitutive" (Reisigl & Wodak, 2009: 89) and are therefore crucial to the way knowledge of social reality is constructed. As conspiracy theories become increasingly popular, the importance of discourse can only be perceived as crucial, as "dissemination to large audiences enhances the constitutive effect of discourse – its power, that is, to shape widely shared constructions of reality" (Mautner, 2008: 32).

The construction of the epistemic masquerade inherent in conspiracy discourses is facilitated by two primary discursive strategies: manipulation and legitimation. These strategies are well-established, having long been employed to analyse the mechanisms of persuasion, particularly within political and media discourse. Far from being an exclusively conspiratorial strategy, manipulation is a discursive strategy typical of language *per se*, especially in the set of discourses used to exercise social domination (e.g. news discourse, political discourse). According to van Dijk, "manipulation, socially speaking, is a discursive form of elite power reproduction that is against the best interests of dominated groups and (re)produces social inequality" (van Dijk, 2006: 364). van Dijk's famous definition was originally conceived to explain the way manipulation was deployed by the dominant classes to rule over the masses and maintain social hegemony. However, manipulation is also employed by all those who wish to achieve social hegemony in direct opposition to those who already exert it. van Dijk argues that there are four strategies of manipulation in discourse: (1) the speaker's claim to credibility and moral superiority in opposition to the opponent's lack thereof; (2) the authenticity of the speaker's beliefs and the evidence which is brought to testify to that; (3) the speaker's attacks on his/her opponent and his/her beliefs, for example through the "ideological square", i.e., the discursive construction of an *us vs them* dichotomy, whereby an us-group is given positive qualities and a them-group is given negative qualities; (4) appealing to the emotions, ideas and attitudes of the recipient (van Dijk, 1998, 2006). When these strategies are deployed, the description of certain events or people becomes partial, deceiving and, at least partly, potentially falsifiable, as certain aspects are obfuscated while others are highlighted. The ideological square is crucial in the polarized representation of conflict between two different groups, often identified in discourse with the pronouns we/us and they/them. These pronouns do not merely fulfil a grammatical function but are powerful indicators of collective belonging/solidarity/inclusion or estrangement/othering/exclusion. By using *we* and *them* as opposites and as coherent, consistent social wholes, speakers simplify social identities and reduce "complex social variability to a single dimension: us versus them" (Bucholtz & Hall, 2005: 384). All the while, the use of *we* can also promote the idea of consensus, thus manipulating people's perception of reality.

Legitimation is achieved by using certain linguistic and non-linguistic instruments to validate statements or actions or to transform subjective stances into objective ones – in other words, this happens when "a system of linguistic objectification of human experience is transmitted" (Berger & Luckmann, 1966: 112), and this is done in order to persuade audiences. According to van Leeuwen and Wodak (van Leeuwen, 2007; van Leeuwen & Wodak, 1999), there are four categories of legitimation: (1) Authorization, or the use of traditional or legal authority; (2) Moral Evaluation, whereby legitimation occurs through reference to value systems; (3) Rationalization, or the reference to institutionalized social actors who possess cognitive validity); and (4) Mythopoesis, that is, the use of storytelling. These strategies are exploited in all kinds of language use but become crucial in politics and news, which are discursive realms with a strong social impact and where persuasion plays a key role. Authorization, in particular, is a key strategy of legitimation as it employs the authority of other sources to legitimize the speaker's point of view. Within Authorization, van Leeuwen operates a further distinction, identifying Personal Authority, Expert Authority, Role Model Authority, Impersonal Authority, and the Authority of Tradition. Expert Authority and the Authority of Tradition are rendered concrete in language especially through quotation, an intertextual feature whereby parts or wholes of text are inserted into the text in question. The purpose of quotations may be multiple, as they can be used to delegitimize their original sources, or they can be used as sources of authority to bolster what is being said or written. Quotations have an epistemological purpose, as they provide discourse with evidentiality through statement of facts. News stories, in particular, are constituted by the integration of materials coming from different sources (e.g. videos, interviews, official reports, stories from social media). In this way, credibility and objectivity are communicated not as coming from the writers' own voice, but as validated by evidence and facts coming from "external" authorities (Bednarek & Caple, 2012: 90–94).

As discussed in my previous work (Demata, 2024), the multimodal texts which populate the digital environment are rife with quotations and references to other texts, both verbal and non-verbal. Photos, videos, memes, links to external websites and retweets function as recontextualized and shared contents and are in themselves forms of quotation. The circulation of contents through quotation and sharing ensures visibility and aims at legitimation: sharing contents is a means for users to engage with each other and to legitimize each other's opinions (or to delegitimize those of others). Quotations can also have a political purpose and their analysis is crucial when assessing the way certain social values and hierarchies are encoded in discourse and the manner in which what is described is being recontextualized (Fairclough, 1992, 1993; Krzyżanowski, 2016). In line with van Dijk's second strategy of manipulation, i.e. focusing on the authenticity of the speaker's beliefs and evidence, quotations may also be used as instruments of manipulation, as they allegedly provide speakers with evidence that confirms their point of view and discredits that of their opponents.

Manipulation as an instrument of truth-building is also achieved due to the structure of the modern media environment itself. Manufacturers of media disinformation and conspiracy theories rely on the fact that contents can go from platform to platform basically unchecked and unchallenged, thus achieving legitimation and credibility by virtue of their being shared and made visible through different sources. Information can be decontextualized and recontextualized in contexts of production and use which are increasingly decentralized, and therefore more prone to the actions of disinformation agents. What happens with disinformation and conspiracy theories is the formation of "evidence collages" which filter out inconvenient or hostile claims and support certain (otherwise potentially refutable) claims (Kraft & Donovan, 2020).

6.4 The WHO and the conspiracy theories about the pandemic

Founded in 1948 as an international institution directly dependent on the United Nations, the WHO has the task of coordinating international efforts to tackle health issues, assisting governments in matters of health and safety, especially transmittable diseases. The WHO has often been a target for conspiracy theorists. For example, it was claimed the CIA had created HIV and the WHO had been instrumental in spreading it through polio vaccinations in Africa. This was supposedly done in order to reduce the world population (Kalichman, 2009). The WHO has often been seen as the perfect embodiment of establishment science and, accordingly, as an instrument in the hands of the United Nations or of more or less hidden powers conspiring against the health of citizens as well as their freedom. These beliefs intensified with COVID-19, whose exact origins and nature were to remain (and are still) largely unknown. Political and health institutions worldwide were engulfed in forms of crisis communication which did little to combat the general uncertainty about the virus that was perceived on a global scale (Wodak, 2021). The WHO itself was perfectly aware of the problems that a new and seemingly unstoppable virus could create in the world of news and information. In a report published in February 2020, they coined a new term, "infodemic", which rapidly gained widespread popularity:

> The 2019-nCoV outbreak and response has been accompanied by a massive 'infodemic' – an over-abundance of information – some accurate and some not – that makes it hard for people to find trustworthy sources and reliable guidance when they need it.
>
> (WHO 2020)

Truth be told, during the pandemic the WHO did not always provide univocal and reliable information. From early 2020, when the pandemic broke out, the WHO began publishing a series of statements on the origins and spread of the virus which they would later retract or revise. For instance, the WHO first claimed that the virus could not be airborne, only to then retract such claims (Molteni, 2021). This hesitant behaviour gave origin to all sorts of conspiracy narratives whereby, for example, the WHO was supporting Chinese interests by covering up the supposedly lab-made origins of the virus.

The uncertain knowledge about the origins and spread of the pandemic and the often vague communication regarding its containment from scientific and political institutions worldwide were two key contributing factors to the rapid growth and popularity of anti-science and conspiracy narratives. These narratives cast doubts on the existence of the virus itself, and later of the effectiveness of vaccinations to prevent the disease (Constantinou, Kagialis, & Karekla, 2021; Vâlcea, 2023). The uncertainty regarding Expert Authority and the ensuing spread of fake news and disinformation regarding the pandemic and the anti-COVID vaccines were certainly a global phenomenon taking place in many nations, as illustrated by Larsen and Roslyng's (2025) chapter addressing the case of Denmark in this volume. In the USA, political culture has traditionally been ripe with conspiracy theories, and generalized scepticism towards scientific expertise and communication has a well-established history (Goldberg, 2001; Barkun, 2013; Uscinski & Parent, 2014). However, the level of diffusion of conspiracy narratives about the pandemic was unprecedented compared to earlier periods, no doubt induced by the new media environment, and specifically by the pervasiveness of social media. Conspiracy theories about the virus and the vaccines that were meant to combat it enjoyed huge popularity and diffusion. The life of these conspiracy theories could be traced as emerging from either social media or fringe websites, or both, only to then achieve mainstream status, usually through their being peddled by mainstream politicians or celebrities (Blevins et al., 2021; Bruns, Harrington, & Hurcombe, 2021). These conspiracy narratives targeted all aspects of the pandemic and offered a wealth of different explanations, including the origins of the pandemic as the result of a lab leak, its connection to the 5-G mobile technology, the use of vaccines both in terms of their impact on public health and their potential role as a mechanism for population control and the restriction of democratic freedoms. Similarly, the COVID-19 pandemic was interpreted by some as a tool for advancing the so-called Great Reset, whereby migrant populations were being deliberately positioned to replace white populations over time (Birchall & Knight, 2023). Large sectors of society both in the USA and the rest of the world were deeply influenced by these counter-narratives and conspiracy theories, with the result that the health of many people was severely endangered, lives were lost, and faith in the political process undermined.

6.5 Alex Jones and InfoWars

Alex Jones fully exploited the feeling of mistrust in public health organizations and political institutions during the pandemic. He was, in fact, a major contributor to the spreading of disinformation about COVID-19. Jones is, without any doubt, one of the world's most famous conspiracy theorists. Through his own platform *InfoWars* and his substantial presence on social media, Jones has always been a highly controversial figure due to his promotion of conspiracy theories. Among the most infamous theories he has endorsed is the claim that the 9/11 attacks were not perpetrated by Al-Qaeda but were instead secretly orchestrated by then-U.S. President George W. Bush. He has also propagated the New World Order conspiracy theory, which posits that a clandestine elite is actively working to curtail individual freedoms and establish a dictatorial global government. Additionally, he has advanced the claim that the Sandy Hook Elementary School mass shooting was a fabrication, a so-called "false flag" operation allegedly staged by advocates of stricter gun control measures (Beauchamp, 2016). His public support of this last conspiracy cost him a trial for defamation from the parents of the victims of the school shooting and he eventually filed for bankruptcy.

Jones has always had a tumultuous relationship with social media: he was banned from Facebook and other social media in August 2018 (Riley, 2018) and again in May 2019 (Darcy, 2019) for spreading hate speech and openly calling for violence. For the same reasons, Apple also deplatformed Jones in 2019 and Google Play Store followed suit in May 2020 because of the amount of disinformation he was spreading about the pandemic. Indeed, the InfoWars Facebook page and other Facebook pages related to it endorsed anti-vaccination views (Zadrozny, 2019). Jones compared coronavirus lockdowns to "what Nazis did" (Knox, 2020). Through his networks, he sold fake supplements (a considerable source of income for him) and an anti-COVID toothpaste which were allegedly treatments against the virus and were banned from commerce (Ferre-Sandurni & McKinley, 2020; Marantz, 2020; Porter, 2020). Jones certainly had a rather extensive following, and InfoWars itself, as a platform, often competed with, and was more successful than, mainstream websites such as Newsweek or The Economist (Beauchamp, 2016).

Jones's attacks on the WHO form part of his general view of a global conspiracy, in which the pandemic and vaccinations are all instruments used by an all-powerful elite to gain authoritarian global world domination. Jones's prominence is largely attributable to his presence in online videos, which are published on his own InfoWars website and disseminated across various digital platforms and social media. Consequently, a thorough analysis of his conspiratorial discourse must consider the ways in which his online videos construct "truth" through a combination of verbal and non-verbal strategies.

6.6 The dataset

The dataset used in this chapter consists of two videos. The first video was published on the "Alex Jones" channel on Rumble (https://rumble.com/c/c-618 485) on 7 May 2022 and is entitled "BREAKING Alex Jones Warns The World Health Organization Is Attempting a Power Grab!" (https://rumble.com/v13uuor-breaking-alex-jones-warns-the-world-health-organization-is-attempting-a-pow.html). *Rumble* (www.rumble.com) is an online video platform, initially conceived as an alternative to YouTube; it routinely hosts far-right and Republican politicians as well as conspiratorial contents. Jones's short video is taken from InfoWars itself and was found by inserting "Alex Jones" and "World Health Organization" into the Rumble search engine. It was chosen because it was the most widely shared and reposted by several other pages on Rumble itself.

The second video was extrapolated from a page on the InfoWars website focusing on the statements made by biologist Bret Weinstein about the WHO's alleged intention to introduce new health plans which would threaten democracy and free speech. The page, entitled "Biologist Warns WHO Planning to Take Control Over Nations For Next Pandemic, Confirms COVID "Vaccines" Killed 17 MILLION People" (www.infowars.com/posts/biologist-warns-who-planning-to-take-control-over-nations-for-next-pandemic-confirms-covid-vaccines-killed-17-million-people/), appeared on 6 January 2024 and was selected for this research because it is the first of a series of articles appearing in InfoWars in which Jones, in the following two weeks, discussed Weinstein's claims at length. The analysis will focus on Jones's video which appears at the bottom of the page, before the viewers' comments, and is entitled "Experts Claim Global Heart Failure Pandemic Caused by New COVID Strain", in which he elaborates on Weinstein's interview. As is the case for all webpages, the text consists of a combination of written text, photos and videos. Weinstein's original statements were released during an episode of "The Tucker Carlson Encounter", a program aired through the Tucker Carlson Network, in which its anchor interviewed Weinstein. Carlson is a very controversial radical right-wing anchorman and a supporter of conspiracy theories. He was fired from Fox News in April 2023 and founded his own streaming service in December 2023.

6.7 Analysis

In both videos analysed in this study, as in all content produced for InfoWars, Jones adopts the demeanor of a broadcaster in a professional newsroom, presenting against a backdrop featuring a large screen displaying digital world maps. These videos are highly polished in terms of production quality and employ visual settings that emulate those of established news organizations. Consequently, this carefully curated environment provides viewers with a sense of familiarity and credibility, creating the impression that they are engaging with an authentic and legitimate news program.

In "BREAKING Alex Jones Warns The World Health Organization Is Attempting a Power Grab!", Jones argues that the WHO is going to pass amendments to the IHR, the International Health Regulations, at an upcoming meeting of the World Health Assembly taking place in Geneva on 22–28 May 2022. These amendments would then be sent to, and approved by, the US Senate and, as a consequence, according to Jones, the US would be further deprived of its sovereignty. The video starts by addressing the broader frame within which the amendments are to be approved:

> As you know, the United Nations was set up by the Rockefeller Foundation and other multinational billionaire companies after World War II. It's always been trying to set up a planetary world government that controls food, the medicine, travel, reproduction, everything towards a planned economy that depopulates the Earth.
>
> (0:07–0:24)

Jones's words immediately place the claims about the proposed amendments to the IHR within the frame of the New World Order conspiracy theory. As discussed in Section 6.4 of this chapter, the WHO is perceived as a tool employed by the United Nations to establish global dominance in accordance with the objectives set forth by the New World Order. This objective would purportedly be realized through the implementation of various restrictive health regulations, including global vaccination, a premise that Jones uncritically assumes to be true. Indeed, the incipit of Jones's monologue, "As you know", presupposes a common knowledge of facts shared by Jones and his audience and a mutually agreed upon evaluation of them. It is an evidential claim based on presupposition and aims to establish an immediate connection with his viewers.

The conspiratorial notion whereby the US Senate was willing and ready to ratify the amendments of the IHR is presented as truth by legitimizing it through a strategy of Authorization. This is achieved at two levels. The first is the purely linguistic one, based mainly on Jones's narrating voice. He constructs legitimacy through intertextuality as he reads extracts from an article and a webpage which are shown on the screen:

> Here's a few articles, and they break it all down. "Bill Gates Wants to Build a New Global Team Called GERM," it's this whole UN United Nations treaty that goes over it all, we'll be covering it live on air, but this is all unfolding right now in Geneva, Switzerland the 27th and the 28th excuse me… so there it is, it's a great site that breaks it down… DontYouDare.INFO they did a great job kind of putting it all together.

The articles shared by Jones with his viewers are from two hyperpartisan websites, Gateway Pundit (www.thegatewaypundit.com) and DontYouDare.info. Jones uses these articles as evidence of the WHO's conspiracy to achieve a "power grab". The article in the Gateway Pundit is copied and pasted from ProTrumpNews.com, another hyperpartisan website. This section of Jones's video therefore includes a quotation of a quotation. Quoting these articles means using Authorization, whereby sources considered reputable confirm the speaker's point of view. Furthermore, it is not just the cited text that confers Authorization, but the *act* itself of quoting and sharing a certain source somehow amplifies certain claims and bestows legitimacy upon them, especially when this is done multiple times, as in the case of the article from ProTrumpNews. As discussed in Section 6.4, this is a strategy used by legitimate journalism when quoting external sources or other news outlets in order to lend authority to what is claimed.

At 1:01–1:03, the camera shows a section of the article published in Gateway Pundit:

> The 75th meeting of the World Health Assembly will be held in Geneva, Switzerland this May 22–28, 2022. The Assembly will vote on the amendments to the IHR. They are very likely to pass and be enacted into international law unless "We the People" stand up against this attack on our sovereignty.

The mention of "We the People" combines two strategies: (1) the ideological square, which consists of two groups, us vs them, who are constructed in discourse in opposition to each other, through positive self-presentation and negative other-presentation. This is a typical populist/conspiracist strategy, whereby the "us" is construed in opposition to the "them", i.e. the elite or the establishment, and the two groups are seen as perfectly consistent; (2) the Authority of Tradition, as the mention of "We the People" is a reference to the initial words of the Preamble to the United States Constitution, which is taken to be common knowledge in US society, and constitutes the evaluative force of Jones's argument. Through the combination of these two strategies, Jones aligns his opinions with those of "we the people". He grants credence to his claims by making use of the highest authority in the country, as the words from the Constitution legitimize his own stance in opposition to those of the conspiratorial elite.

The second level in which truth is constructed through Authorization in the video comes from Jones's physical environment. His quotations from the two websites are followed by him pointing to his desk and saying: "I've got hundreds of articles over here we are covering right now", with the camera showing several small piles of sheets of paper which are meant to confirm Jones's claims (Figure 6.1). Authority through textual reference is therefore realized via non-textual elements, i.e. the visual elements pointed out by Jones in the video. We might call this presupposed intertextuality, as it is assumed that what Jones is saying about the WHO's conspiratorial planning is validated by other texts, i.e. those spread across the desk, which viewers do not have direct access to but which Jones considers as validating sources of authority. The authority in this case is asserted not on the basis of the supposed contents of the articles (which viewers cannot read and therefore do not have knowledge of), but on the basis of their quantity ("I've got hundreds of articles") and their presence and visibility on Jones's desk.

The video ends with, once again, the assumption that there is a conspiracy going on and a call to action to his viewers, who are encouraged to share the news he is airing:

> The pandemic was just phase one. This is phase two. We got to maintain America's sovereignty. God bless, and good luck, now get started, get this information out please.

Figure 6.1 Screenshot from "BREAKING Alex Jones Warns The World Health Organization Is Attempting a Power Grab!", 7 May 2022 (https://rumble.com/v13uuor-breaking-alex-jones-warns-the-world-health-organization-is-attempting-a-pow.html)

Thus, the narration of the video shifts from reporting the news from other websites and platforms, used as sources of authority, to evidential statements about a conspiracy, whose existence is taken for granted. This communicates an impression of veracity to what Jones says.

The second video under analysis, "Experts Claim Global Heart Failure Pandemic Caused by New COVID Strain", is included as part of a page in the InfoWars webpage entitled "Biologist Warns WHO Planning to Take Control Over Nations For Next Pandemic, Confirms COVID "Vaccines" Killed 17 MILLION People". In the page, Jones discusses an interview in which Bret Weinstein stated that the COVID vaccine caused extensive damage to people's health worldwide. Weinstein also claimed that the resistance of a "small group of dissidents" against the vaccines managed to somehow alter the dominant narrative and highlight the dangers of the WHO's plans to both democracy and human health.

As with all webpages, the article which appears on InfoWars is constructed as an ensemble of verbal and visual texts. The verbal sections consist mainly of extracts from Weinstein's interview. They are interspersed with four tweets, each featuring video extracts from Weinstein's interview for "The Tucker Carlson Encounter", embedded in the webpage, and with several links to the biologist's full interview. As in the case of the first video, information is communicated and legitimized through the intertextual presence of external sources. In the title of the article, the mention of the fact that Weinstein is a biologist imbues the whole narrative with authority. Following the strategy of Authorization, and specifically that of Expert Authority, what is said about the pandemic is taken as true and legitimate because these statements are made by a reputable man of science.

Jones's own 7:19 video was originally published in Banned Video, another platform controlled by Jones himself. Once again, Jones is in the InfoWars newsroom commenting and expanding upon the news presented in the page. The WHO is not directly mentioned in the video but was implicitly held responsible both for the virus' effects on people's health and for the restrictive measures that were introduced in order to contain the spread of COVID-19.

As in the first video analysed previously, Jones is shown in his own newsroom sitting at a desk strewn with copies of news articles (Figure 6.2). The visual aspects of the video aim once more to communicate the abundance of sources available to Jones, something that legitimizes his claims. In the video, Jones makes several appeals to the scientific evidence contained in the articles he reads. Authority is achieved by quoting Weinstein's 17-million claim and merging it with other scientific sources and data:

> *We have top scientists, with the nations that have reporting data, so it's higher than that, 17,000,000 conservatively killed by the shots.* The RAS [renin-angiotensin system] infected with prion crystals that basically create cancer and cut the DNA strand right at the key spot. I mean, they have hit us and hit us hard, with a soft kill delayed reaction weapon. Now I'm sorry to have to announce

Figure 6.2 Screenshot from video featured in "Biologist Warns WHO Planning to Take Control Over Nations For Next Pandemic, Confirms COVID 'Vaccines' Killed 17 MILLION People", 6 January 2024 (www.infowars.com/posts/biologist-warns-who-planning-to-take-control-over-nations-for-next-pandemic-confirms-covid-vaccines-killed-17-million-people/)

> that to you, but that's why the illegal aliens are exempt, they want them as the new replacement population. But shit, shit, shit, it's just the reality and it's horrifying. Here's some more of the news on that front. And then we're going to your phone calls.[1]

The mention of "We have top scientists", again following Expert Authority, lends an air of credibility to the whole statement by Jones, who then goes on to use complex medical language ("The RAS infected with prion crystals that basically create cancer and cut the DNA strand right at the key spot") that goes well beyond basic layman knowledge, thus creating a layer of authority through obfuscation, i.e. the use of a very specialized language known to experts only. This is followed by an explanation of the reason why migrants may allegedly opt out of the vaccination campaign ("that's why the illegal aliens are exempt, they want them as the new replacement population"). This last statement places Jones's supposedly empirical explanation in the context of the Great Reset conspiracy theory.

Jones then shows the audience a number of articles taken from various sources and reads some headlines and short extracts from them. The articles originate from various hyperpartisan news outlets and platforms, including Breitbart, ZeroHedge, Gateway Pundit, and RedState, as well as InfoWars itself on three separate occasions. Jones presents these sources as substantiating his claims regarding vaccines, as each of them cites statements attributed to purported medical authorities and experts:

> Experts say new COVID strain will cause global heart failure pandemic. Global heart failure Pandemic. And they go on and say ohh *scientists around the world and in Japan* they've they've looked at the new variant, it has a spike protein on it – they all do, the HIV spike protein – and it attacks the heart and causes viral myocarditis.
>
> Here's a Fox News Report: "*Doctor Mark Siegel* on what's behind the decline in U.S. life expectancy" – hmm mmm for the first time ever – *"up 25% among 35 to 44 year-olds and up 19% among 25 to 34 year-olds*, but the vaccine which supposedly saved lives, nobody is studying what the long term effects of the vaccine here are". That's also an elephant in the room. Didn't save any lives, it was all lies.
>
> Yep. *New study*: "Safety Concerns about the Covid Shot Arise", "New Safety Concerns about the Covid Shot Arise".

Again, following Expert Authority, credibility is lent by the authoritativeness of the people quoted ("Experts", "Doctor Mark Siegel"), by the fact that his quotes are part of some scientific report ("New study"), and also because the sources of these quotes are external authorities, or are presented as such (Zerohedge, Fox News, and InfoWars respectively). As in the case of the conspiratorial claim whereby "illegal aliens" are going to become "the new replacement population", at the end of the Fox News Report Jones makes another typical claim, "That's also an elephant in the room. Didn't save any lives, it was all lies," again mixing the seemingly scientific evidence with his own personal assessment of the situation.

At the end of the video, after having finished reading the headlines of some of the articles spread across his desk describing the supposedly adverse effects of vaccines, Jones exclaims: "That's quite a stack of news on that front, isn't it". Again, just like the "hundreds of articles" at the end of the "BREAKING Alex Jones Warns The World Health Organization Is Attempting a Power Grab!" video, the supposedly high number of sources and their visibility to the viewers have a cumulative effect in terms of legitimation, as if the validation of Jones's claims were due to the sheer quantity of articles displayed by him to his viewers.

6.8 Discussion

The discursive strategies in Jones's videos analysed in this chapter are the textual foundation of the epistemic masquerade which I consider to be the essence of conspiratorial discourse. The construction of truth, which is negotiated through the three dimensions of masquerading – ideological, populist and communicative – are conveyed in the language used by Jones by means of strategies of manipulation and legitimation. Through these strategies, an epistemology of knowledge, itself the basis of social and personal conduct, is upheld, however "false" or untrue it may be. Crucially, the conspiracy counter-discourse is masqueraded as truth by using certain visual and verbal strategies that mimic those of the official discourses of epistemological authorities: Jones communicates contents which are clearly oppositional and anti-establishment but are well *masqueraded* as meaningful, rational and legitimate as they recall familiar visual frames, terms of reference and modes of speech.

First of all, Jones's conspiracy discourse occurs within a visual setting which is recognizable. As has been mentioned in the previous section of this chapter, the newsroom environment at InfoWars, with Jones as its anchorman, looks authentic and legitimate and recalls that of traditional newsrooms. In other words, in order to claim credibility, Jones borrows the visual conventions of mainstream media. Interviews and other sources are used in much the same way as legitimate journalism. The details of the visual setting, with the "stacks or articles" strewn across Jones's desk for his viewers to see, also add elements of tangibility and veracity to his claims.

Second, at the purely textual level, Jones masquerades his claims by employing discursive strategies that are very frequent in traditional or establishment political and media discourse, and generally in those discourses that aim at persuasion in the public sphere. van Dijk's manipulating strategy of the "ideological square" is the discursive expression of Jones's Manichean worldview, a worldview typical of conspiracy theorists, whereby the world is seen as a battlefield between the forces of good and those of evil. The ideological square constructs an opposition between us and them in discourse: the us-group is granted the status of morally good actors, while the members of the them-group are constantly represented as evil-doers. In Jones's videos, this polarized positioning is embodied discursively in the people, or the USA, vs the United Nations, the WHO, Bill Gates, and whoever else conspires to the detriment of people's health and freedom. Jones's enemies have also been the frequent targets of populist leaders and movements for pretty much the same reasons: the WHO, the United Nations and other obscure actors manufactured the virus or exaggerated its impact, hid information from the people, and have plans to subjugate mankind through the vaccine, either directly through the vaccine's effects or indirectly by establishing a "power grab" at the expense of the popular will.

The epistemic masquerade of conspiracy theories also takes place at the discursive level through strategies of legitimation, by which conspiracy discourse both mimics and opposes the establishment discourse of science. Thus, an alternative system of knowledge is constructed, functioning as an epistemological framework, a value system and a counterpoint to the hegemonic paradigm upheld by the establishment. Indeed, while Jones actively critiques and seeks to dismantle the official scientific discourse of the WHO, his own discourse continues to be framed, and even deliberately *masqueraded*, as scientific, and it incorporates counter-evidential claims which he purports to be *true*. Jones's discussion of the WHO's plans for a "power grab" and of Weinstein's interview regarding the dangers of vaccines is rife with strategies of legitimation, such as Authorization, which manage to transform his manipulated, partial or pseudo-scientific claims into what appears to be truth.

Jones utters authentically conspiratorial statements during presentations which are replete with supposedly authoritative statements issued by scientists. Thus, legitimation stems from the use of sources which are apparently empirical, i.e. medical and scientific reports, quoted from and mediated by recognizable media, i.e. news websites and platforms. Indeed, credibility is achieved through quoting, sharing, and referring to external sources multiple times, these sources being presented as "evidence collages" claiming legitimacy by means of intertextual chains.

6.9 Conclusion

In the conspiratorial mind set, "official" science is attacked not just because of its supposed fallacies or manipulations. Supporters of conspiracy theories see science as a key segment within a socially encompassing worldview: scientists are attacked for the social and political authority and role they hold, not just for the (supposedly wrong) scientific ethos they support. During the pandemic, the WHO played a key role in addressing the crisis within the international public sphere. Not surprisingly, the WHO also became a frequent target for conspiracy theorists and often the focal point for conspiracy theories regarding COVID-19 and the vaccines created to combat the virus. Given the importance of science in the public debate at the time of the pandemic, conspiracy theories regarding the broad spheres of health and science became very popular, attacking, as they did, the foundations of the official narratives issued by the epistemological authorities.

In order to claim legitimacy and present itself as "true", conspiracy discourse is aligned to (and makes use of) certain strategies which are typical of scientific and news discourses. Alex Jones's videos analysed here mimic the reliability and trustworthiness of "legitimate" news in more than one way and from more than one semiotic perspective. The claim to truth of conspiracy theories has often been characterized as manipulative, false and deceiving: the standard view of conspiracy is indeed one based on the absurdity of the claims advanced by conspiracy followers, stemming from the assumption that all that they state as fact is false. However, persuasion in Jones's discourse is achieved through the presentation of a mixture of conspiratorial beliefs and traditional discursive strategies based on (allegedly) empirical evidence. This presentation lends authority and credibility to what he says. The New World Order, the Great Reset and other conspiracies are based on epistemic masquerade, whereby conspiratorial claims are legitimized by scientific (pseudo-)evidence masquerading as truth. Indeed, it may be said that thanks to a number of well-honed discursive structures, falsehoods are disguised or masqueraded and thus granted an aura of authenticity which may well be (mis) taken for truth.

Note

1 Emphasis in this quotation and in the following ones is by the author.

References

Barkun, M. (2013). *A culture of conspiracy*. (2nd ed). Berkeley: University of California Press.

Beauchamp, Z. (2016). Alex Jones, Pizzagate booster and America's most famous conspiracy theorist, explained. *Vox*. 7 December 2016. www.vox.com/policy-and-politics/2016/10/28/13424848/alex-jones-infowars-prisonplanet

Bednarek, M. & Caple, H. (2012). *News discourse*. London: Continuum.

Berger, P.L., & Luckmann, T. (1966). *The social construction of reality: A treatise in the sociology of knowledge*. New York: Doubleday & Company.

Bergmann, E. (2018). *Conspiracy and populism: the politics of misinformation*. London: Palgrave.

Bhatia, A. (2015). *Discursive illusions in public discourse*. London: Routledge.

Birchall, C., & Knight, P. (2023). A perfect storm. Covid-19 conspiracy theories in the United States. In M. Butter & P. Knight (Eds.), *Covid conspiracy theories in global perspective* (pp. 335–350). London: Routledge.

Blevins, J.L., Edgerton, E., Jason, D.P., & Lee, J.J. (2021). Shouting into the wind: medical science versus "B.S." in the twitter maelstrom of politics and misinformation about hydroxychloroquine. *Social Media + Society 7*(2) https://doi.org/10.1177/20563051211024977

Blommaert, J. (2020). Political discourse in post-digital societies. *Trabalhos em Linguística Aplicada*, *59*(1), 390–403. https://doi.org/10.1590/01031813684701620200408

Bruns, A., Harrington, S., & Hurcombe, E. (2021). Coronavirus conspiracy theories: tracing misinformation trajectories from the fringes to the mainstream." In M. Lewis, E. Govender & K. Holland (Eds.), *Communicating COVID-19: interdisciplinary perspectives* (pp. 229–249). Cham, Switzerland: Springer,.

Bruns, A., Hurcombe, E., & Harrington, S. (2022). Covering conspiracy: approaches to reporting the COVID/5G conspiracy theory. *Digital Journalism 10*(6), 930–951. https://doi.org/10.1080/21670811.2021.1968921

Bucholtz, M., & Hall, K. (2010). Locating identity in language. In C. Llamas & D. Watt (Eds.), *Language and identities* (pp. 18–28). Edinburgh: Edinburgh University Press.

Chadwick, A. (2017). *The hybrid media system: politics and power* (2nd ed). Oxford: Oxford University Press.

Chapelan, A. (2021). "Swallowing the red pill": the coronavirus pandemic and the political imaginary of stigmatized knowledge in the discourse of the far-right." *Journal of Transatlantic Studies 19*, 282–312. https://doi.org/10.1057/s42738-021-00073-2

Constantinou, M., Kagialis, A., & Karekla, M. (2021). "COVID-19 scientific facts vs. conspiracy theories: is science failing to pass its message?" *International Journal of Environmental Research and Public Health*. 18/6343. https://doi.org/10.3390/ijerph18126343

Dahlgren, P. (2018). "Media, knowledge and trust: the deepening epistemic crisis of democracy." *Javnost—The Public 25*(1–2), 20–27. https://doi.org/10.1080/13183222.2018.1418819

Darcy, O. (2019) "Louis Farrakhan, Alex Jones and other 'dangerous' voices banned by Facebook and Instagram." *CNN*, 3 May 2019. https://edition.cnn.com/2019/05/02/tech/facebook-ban-louis-farrakhan-infowars-alex-jones-milo-laura-loomer/index.html

Davis, M. (2021). "The online anti-public sphere." *European Journal of Cultural Studies*, *24*(1), 143–159. https://doi.org/10.1177/1367549420902799

Demata, M. (2024). The legitimation of conspiracy theories through manipulation. The case of climate lockdowns. In C. Ilie (Ed.), *Manufacturing dissent: manipulation and counter-manipulation in times of crisis* (pp. 157–183). Amsterdam: John Benjamins.

Demata, M., Zorzi, V., & Zottola, A. (2022), Conspiracy theory discourses: critical inquiries into the language of anti-science, post-trutherism, mis/disinformation and alternative media. In M. Demata, V. Zorzi & A. Zottola (Eds.), *Conspiracy theory discourses* (pp. 1–22). Amsterdam: John Benjamins.

Douglas, K.M., Uscinski, J.E., Sutton, R.M., Cichocka, A., Nefes, T., Ang, C.S., & Deravi, F. (2019). Understanding conspiracy theories. *Advances in Political Psychology 40*(1), 1–34. https://doi.org/10.1111/pops.12568

Enli, G. (2024). Populism as "Truth": how mediated authenticity strengthens the populist message. *The International Journal of Press/Politics* Online first, 1–17. https://doi.org/10.1177/19401612231221802

Fairclough, N. (1992). Intertextuality in critical discourse analysis. *Linguistics and Education* 4/3–4, 269–293. https://doi.org/10.1016/0898-5898(92)90004-G.

Fairclough, N. (1993). *Discourse and social change*. London: Wiley.

Fenster, M. (2008). *Conspiracy theories: secrecy and power in American culture* (Revised ed). Minneapolis: University of Minnesota Press.

Ferre-Sandurni, L.F., & McKinley, J. (2020). Alex Jones is told to stop selling sham anti-coronavirus toothpaste. *The New York Times*, 13 March 2020 www.nytimes.com/2020/03/13/nyregion/alex-jones-coronavirus-cure.html

Fuchs, C. (2021). *Communicating COVID-19*. Bingley: Emerald.

Goldberg, R.A. (2001). *Enemies within: the culture of conspiracy in modern America*. New Haven, CT: Yale University Press.

Harambam, J. (2020). Contemporary conspiracy culture. truth and knowledge in an era of epistemic instability. London and New York: Routledge. https://doi.org/10.4324/9780429327605

Harambam, J., & Aupers, S. (2015). Contesting epistemic authority: conspiracy theories on the boundaries of science. *Public Understanding of Science 24/4*, 466–80. https://doi.org/10.1177/0963662514559891

Kalichman, S. (2009). *Denying AIDS: conspiracy theories, pseudoscience, and human*. Göttingen: Copernicus-Springer.

Knox, P. (2020). Controversial Infowars conspiracy theorist Alex Jones compares coronavirus lockdowns to 'what Nazis did'. *The Sun*, 19 April 2020. www.thesun.co.uk/news/11429179/controversial-infowars-conspiracy-theorist-alex-jones-coronavirus-nazi/

Krafft, P.M., & Donovan, J. (2020). Disinformation by design: the use of evidence collages and platform filtering in a media manipulation campaign. *Political Communication 37*(2), 194–214. https://doi.org/10.1080/10584609.2019.1686094

Krzyżanowski, M. (2016). Recontextualisation of neoliberalism and the increasingly conceptual nature of discourse: challenges for critical discourse studies. *Discourse & Society 27*(3), 308–321. https://doi.org/10.1177/0957926516630901

Larsen, G. & Roslyng, M.M. (2025). Knowledge and counter-knowledge: the construction of facts in the vaccination debate. In M.M. Roslyng, A. Rantasila & A.M. Jönsson (Eds.), *Communicating science, climate change and the environment in hybrid media: constructed facts, contested truths* (pp. 112–130). London: Routledge.

Mäkinen, M. (2025). Citizen activists or pandemic deniers? Alternative voices in the Finnish journalistic media during the COVID-19 pandemic. In M.M. Roslyng, A. Rantasila & A.M. Jönsson (Eds.), *Communicating science, climate change and the environment in hybrid media: constructed facts, contested truths* (pp. 131–150). London: Routledge.

Marantz, A. (2020). Alex Jones's Bogus coronavirus cures. *The New Yorker*. 30 March 2020. www.newyorker.com/magazine/2020/04/06/alex-jones-bogus-coronavirus-cures

Mautner, G. (2008). Analysing newspapers, magazines and other print media." In R. Wodak & M. Krzyzanowski (Eds.), *Qualitative discourse analysis in the social sciences* (pp. 30–53). Basingstoke: Palgrave Macmillan.

McIntyre, L. (2018). *Post-truth*. Boston: MIT Press.

Molteni, M. (2021). The 60-year-old scientific screwup that helped covid kill. *Wired*, 13 May 2021. www.wired.com/story/the-teeny-tiny-scientific-screwup-that-helped-covid-kill/

Porter J. (2020). Alex Jones ordered to stop selling fake coronavirus cures. *The Verge*, 13 March 2020 www.theverge.com/2020/3/13/21177998/new-york-attorney-general-alex-jones-infowars-coronavirus-cures-prevention

Reisigl, M., & Wodak, R. (2009). The Discourse-historical Approach (DHA). In R. Wodak & M. Meyer (Eds.), *Methods of critical discourse analysis* (pp. 87–121). London: Sage.

Riley, C. (2018). YouTube, Apple and Facebook remove content from InfoWars and Alex Jones. *CNN*, August 6, 2018. https://money.cnn.com/2018/08/06/technology/facebook-infowars-alex-jones/index.html

Sunstein, C.R., & Vermeule, A. (2008). Conspiracy theories: causes and cures. *Journal of Political Philosophy*, *17*(2), 202–227. https://doi.org/10.1111/j.1467-9760.2008.00325.x

Uscinski, J.E. (2020). *Conspiracy theories. A primer*. London: Rowman & Littlefield.

Uscinski, J.E., & J.M. Parent (Eds.) (2014). *American conspiracy theories*. Oxford: Oxford University Press.

Vâlcea, C. (2023). Anti-science narratives as a form of legitimization of post-truth. *Philologica Jassyensia 19/1*(37), 257–71. https://doi.org/10.60133/PJ.2023.1.18

van Dijk, T.A. (1998). *Ideology: a multidisciplinary approach*. London: Sage.

van Dijk, T.A. (2006). Discourse and manipulation. *Discourse and Society 17*(3), 359–83. https://doi.org/10.1177/0957926506060250

Van Leeuwen, T. (2007). Legitimation in discourse and communication. *Discourse & Communication 1*(1), 91–112. https://doi.org/10.1177/1750481307071986

Van Leeuwen, T., & Wodak, R. (1999). Legitimizing immigration control: a discourse-historical analysis. *Discourse Studies 1*(1), 83–118. https://doi.org/10.1177/14614456 99001001005

WHO (World Health Organization) (2020) Novel Coronavirus(2019-nCoV). Situation Report – 13. 2 February 2020. www.who.int/docs/default-source/coronaviruse/situation-reports/20200202-sitrep-13-ncov-v3.pdf

Wodak, R. (2021). Crisis communication and crisis management during COVID-19. *Global Discourse 11*(3), 329–353. https://doi.org/10.1332/204378921X16100431230102

Zadrozny, B. (2019). Vaccine misinformation and Infowars: researchers wary of Facebook's embrace of "Groups." NBC, 16 February 2019. www.nbcnews.com/tech/tech-news/vacc ine-misinformation-infowars-researchers-wary-facebook-s-embrace-groups-n972286

7 Knowledge and counter-knowledge

The construction of facts in the vaccination debate

Gorm Larsen and Mette Marie Roslyng

Introduction

The Covid-19 pandemic revived public debates on vaccination as governments across the globe introduced anti-pandemic measures, such as vaccination programmes targeting the entire population. While most citizens adhered to these measures, in many countries, a small but vocal minority expressed and acted in protest against what they saw as an abuse of power and deprivation of liberty. In this chapter, we explore the Danish vaccination debate during the Covid-19 pandemic and analyse how arguments for and against vaccination were articulated linguistically and discursively across the media environment by authorities, experts, politicians, citizens, and journalists. Ernesto Laclau's political discourse theory (1990; 2005) allows us to argue that different discursive positions flourish in a hybrid media environment, resulting in antagonistic articulations for or against vaccinations. The vaccination debate thus exemplifies a political struggle to fix a particular meaning as (scientific) truth in a way that either challenges or supports science as a social imaginary (Roslyng & Larsen, 2021). We analyse media representations of a language of power and a counter-language using Mikhail Bakhtin's concepts of the centripetal and the centrifugal – which indicate the forces that unite and centralise, or respectively, open and liberate the verbal-ideological world (Bakhtin, 1981). This leads to the following research questions:

How are scientific facts discursively and linguistically constructed in the vaccination debate in printed news and digital media?

To answer this question, the chapter presents a case study of the Danish media debates on Covid-19 vaccinations during the lockdown period. We examine both articles from the news media and discussions from selected Facebook groups that were highly critical of the Danish government's Covid-19 measures. We conduct a qualitative discourse and linguistic analysis of how hegemonic and counter-hegemonic positions articulate competing truths about vaccine risks while drawing on facts – and ideas – from different and often opposing perspectives. In the contentious field of science communication during the time of the pandemic, we therefore examine whether the different media representations of science can be seen as polarising or if they support a hegemonic authoritative knowledge voice. We thus

DOI: 10.4324/9781003479550-9

hold that while media simultaneously represent science in diverse ways, they also play an active role in constructing scientific discourse.

Theory

Recent studies on the role of media during the Covid-19 pandemic crises emphasise the polarised nature of the Covid-19 coverage across different media. Several quantitative studies suggest that Covid-19 news coverage in mainstream media is characterised by both polarisation and politicisation (Hart et al., 2020; Schmidt, 2023). Furthermore, the news coverage during the pandemic in the US, Canada, and the UK seemed to have a moderate to low level of scientific quality: while right-leaning newspapers were mostly sensationalist, left-leaning news sources focused on misinformation and risks (March et al., 2021). Studies of social media indicate that the epidemic led to a fight over information, an infodemic in which social platforms enabled alternative sources of information and alternative information content (Niemiec, 2020, p. 1; Pool et al., 2021, p. 764; Madvig et al., 2022, p. 4). While digital platforms, in some cases, were used to support the dissemination of the authorities' information to citizens, social media were predominantly a source of disseminating disturbing misinformation, assessed as a threat to public health (Pool et al., 2021, p. 768). During the lockdowns, social media served as a link between people but was also a source of spreading fear and panic (Lelisho et al., 2023). The primary focus in research on communication on social media during the pandemic has thus been on polarisation and the role of misinformation, disinformation, and conspiracy theories (e.g., Jalaludin et al., 2024; Bridgman et al., 2020; Demata, 2025).

The preoccupation with misinformation and disinformation in recent studies indicates that the pandemic crisis is mainly seen through the lenses of post-truth politics to understand how alternative information contradicts and challenges expert knowledge as presented by government authorities. Misinformation and disinformation are understood respectively as "information that is initially presumed to be true but is then later corrected" (Lewandowsky et al., 2017, p. 355) and "deliberately misleading information" (Pool et al., 2021, p. 764), often linked to conspiracy theories and fake news (Thompson, 2008), leading to calls for countermeasures and knowledge management strategies (Bolisani et al., 2021). Misinformation is thus defined in contrast to what can be called expert consensus and must be based on "evidence available at the time" (Vraga & Bode, 2020, p. 136). However, many factors qualify this conceptualisation of misinformation and disinformation, most notably that expert consensus can be disrupted by contestations over expert authority or disagreement within fields of science (Vraga & Bode, 2020, p. 137). Furthermore, being shrouded in uncertainty is part of science at times (Cegarra-Navarro et al., 2022). Apart from the general scientific uncertainty inherent in new research, specific uncertainty related to vaccine risks has been emphasised and discussed in relation to the benefit of vaccines (e.g., Zimmerman et al., 2021). Goldenberg (2021, p. 91ff) argues that the criticism of science and the established system should not be understood as a war against science but as a distancing from scientism, and thus a politicisation of science.

Sismondo (2017) insists that the focus on epistemic democratisation must not be abandoned in the face of post-truth politics. While recent literature has focused on science populism as an often dangerous or problematic flip-side to epistemic democratisation, other authors highlight the democratic problems at stake in silencing alternative science voices (Eslen-Ziya & Giorgi, 2022; De Cleen, 2018; Roslyng, 2022; Makinen, 2025). Raffini and Penalva-Verdú (2022, p. 155) point to the increasing polarisation on the issue of vaccine hesitancy amongst populations, accompanied also by high levels of reciprocal stigmatisation between the opposing groups committed to either trust or distrust in governments' Covid-19 policies. Makinen (2025, this volume) shows how Finnish mainstream media during the Covid-19 pandemic stigmatised sceptical citizens, thus potentially contributing to polarisation during the pandemic. Eslen-Ziya et al. examine the role of hybrid emotional echo chambers in the production of populist counter-knowledge positions. In these chambers, mutual opinions "are formed via emotions (linked to opinions/ideologies/partisan identity) and are shared through both digital network and the social environment around the participant" (Eslen-Ziya et al., 2019, p. 35).

To understand the subversive and counter-hegemonic nature of anti-vaccination positions during the Covid-19 crisis, we argue that a discursive and linguistic approach to alternative knowledge is valuable, as it allows us to acknowledge the interlinking of language, power, and politics when analysing how alternative positions develop in dynamic ways in the new hybrid media environment (Roslyng & Larsen, 2021, p. 895). As shown by Demata (2025, this volume), valuable insights can be gained by studying the discursive strategies that promote science scepticism and conspiracies. We adopt and expand on Ylä-Anttila's definition of counter-knowledge as "contestations of epistemic authority by advocating alternative knowledge authorities" (Ylä-Anttila, 2018, p. 358) that can develop into political strategies directed against mainstream policies (Ylä-Anttila, 2018, p. 357). This definition allows us to see anti-vaccination discourses as counter-hegemonic and political as well as contributions to popularised and mediated scientific discussion. Counter-knowledge can, in some cases, be classified as misinformation, but "most alternative knowledge does not counter knowledge that is in fact (easily) falsifiable [...] nor is counter-knowledge necessarily wrong" (Ylä-Anttila, 2018, p. 361, original emphasis). Our previous research has shown, firstly, how both authoritative and alternative scientific discourse online seemingly draw on a similar unitary language that is, however, manifested in creative and surprising ways in the mediated constructions of truths and facts. Secondly, and similarly, we have argued that the media's discursive construction of counter-knowledge online can be seen primarily as counter-hegemonic political demands that subvert the homogeneity of scientific authoritative discourse (Roslyng & Larsen, 2021, p. 909).

The political-discursive construction of knowledge and counter-knowledge

To analyse the discursive construction of knowledge and counter-knowledge in the media, we use political discourse theory. Following Laclau, we consider scientific discourse as a sedimented form of objectivity; that is to say, a horizon of meaning which presents itself alongside some of society's most persistent collective social imaginaries, such as the project of Enlightenment and the positivist idea of progress, seen "as modes of representation of the very form of fullness [...] located beyond the precariousness and dislocations typical of the world of objects" (Laclau, 1990, p. 64). However, as Laclau also points out, any form of social objectivity, presented as a social imaginary, is also contingent and relies on the successful hegemonic institution of particular myths understood as "a space of representation which bears no relation of continuity with the dominant 'structural objectivity'. Myth is thus a principle of reading of a given situation, whose terms are external to what is representable in the objective spaciality constituted by the given structure" (Laclau, 1990, p. 61). Myth thus creates a new space of representation when social imaginaries are dislocated so that their fullness of representation is disrupted, while at the same time "any objectivity, then, is merely a crystallized myth" (Laclau, 1990, p. 61).

The situation during the pandemic seemed to augment both the government's and the media's reliance on science and to further a collective societal strive, or indeed longing, towards the sense of fullness that science as a social imaginary seemed to promise. However, at the same time, the limits of this objectivity became strikingly visible: while vaccinations appeared to be the long-awaited solution to the risks of the pandemic and were presented as such, we examine how several cracks to its 'full horizon of meaning' appeared during the political process that accompanied the unfolding of the vaccination programme. One such challenge came from the political resistance and the counter-knowledge presented by groups of citizens who expressed their discontent with the Covid-19 policies presented by the government. This meant that the groups perceived the vaccination as riskier than Covid-19 itself, thereby presenting a counter-myth.

The discursive representations of science, as revealed in media debates during the pandemic, thus showed the political nature of the social field in question. For Mouffe, 'the political' is the realm in which the antagonistic and conflictual nature of the social is revealed (Mouffe, 2005, p. 9). Therefore, while science is defined as social horizons of meaning that create a coherent discourse based on facts about, e.g., the risks attached to the Covid-19 vaccine, the political indicates the moment in which these facts are questioned, thus revealing the antagonistic nature of contrasting facts in the vaccination debate. Discourses of counter-knowledge, as defined by Ylä-Anttila (2018, p. 361), have been tied to the construction of knowledge that in some cases seems to adhere to a populist political logic. With Laclau (2005), we understand populism as an antagonistic division of the social space according to two overarching logics: first, the discursive construction of "the people" as an empty signifier, and second, the presence of an antagonistic frontier between the (unfulfilled) demands of the people and a dominant elite. The people, as an empty signifier, will seek to fill the representational gap of meaning that has been vacated in a time of crisis (Laclau, 2005, p. 85–86). Agustín and Nissen (2022) show how anti-restriction movements during the Covid-19 pandemic led to the formation of populist counter-publics.

Laclau's discourse theory allows us to analyse the political logic through which the discourses of knowledge and counter-knowledge in mediated public spheres developed during the Covid-19 pandemic. However, the analysis of the way in which the discursive institution of "the people" and "the elite" relies on the power of language can be fruitfully expanded by turning our attention towards Bakhtin's notions of the authoritative discourse/word and the internally persuasive discourse.

The linguistic-discursive construction of knowledge and counter-knowledge

The authoritative discourse represents a manifestation of a unified language that leads to a centralisation of thinking and the limitations of expression. That is, centripetal forces within language establish a system of norms and rules that "centralize the verbal-ideological world" (Bakhtin, 1981, p. 270). The authoritative discourse represents speech that carries social power and domination by establishing norms, hierarchies, and collective identities with an underlying principle to strengthen and maintain social structures, enforce conformity, and shape public opinion.

In contrast to this, the internally persuasive discourse exerts centrifugal forces within language, where a decentralised and creative language becomes the driving force for development and change. It emphasises the active role of individuals or social groups in constructing meaning, providing a space for creativity, self-expression, and the negotiation of diverse perspectives within a given field (Bakhtin, 1981, p. 272).

It is important to note that the internally persuasive discourse and the authoritative discourse are not absolute opposites, but are integrated with each other, so that one does not exist without the other. For example, the internally persuasive discourse, which allows for individual agency and subversion, cannot escape the influence of authoritative discourse and vice versa.

At first glance, it might appear that the language used by protest groups during the Covid-19 pandemic on social media operates as a centrifugal force, expressing creativity and offering alternative perspectives. However, it can also resemble a homogeneous language with limited diversity in content and expression, lacking internal dialogue. If this is the case, the counter-language does not function (within deliberative democracy) as a threat to authoritative discourse. Instead, it serves to reinforce it.

More comprehensively, Bakhtin's thinking is based on a philosophy of dialogism, which deals with the interrelation of language, meaning, the social, and the ethical. The motto of dialogism is that "nothing is in itself." A dialogic utterance is a response to a potential situation, action, reply, or something else (Bakhtin, 1981, p. 410f; Bakhtin, 1984, p. 183). It requires openness in the situation and towards the other party. Not all dialogues are dialogic; a dialogue becomes dialogic when it involves a relationship and the use of a living language. Through this process, individuals enter a social and ethical space where their words are in relation to the words of others, positioning themselves and others within a relational field.

Within this relational field, two terms are essential: heteroglossia and the other's word. Heteroglossia refers to the coexistence of different languages, dialects, and styles within a single language, e.g., different styles in media. The term highlights the existence of multiple voices and different codes of language usage. Heteroglossia encompasses various dimensions of mixed language, including: (1) the actual use of words, phrases, and sentences that indicate the speaker's attitude towards what is being expressed (elevating, mocking, teasing, ridiculing, consenting); (2) the diverse social voices, including the plurality of genres and the dynamics between them; and (3) the representation of different ideological voices (Bakhtin, 1981, p. 263). This means that the debate about the Covid-19 vaccines is most likely to outline a dynamic group of voices.

For Bakhtin, the other's word is an expression of the semantic residue of other people's words in one's own discourse, whether they are spoken or written. He defines the other's foreign word as any word from someone else or in a different language, that is, any word that does not belong to the self (Bakhtin, 1986, p. 143).

The scientific discourse in both traditional and social media will manifest heteroglossia and the other's words in various ways, serving as analytical tools to delineate the distinct discourses. Probably they will be radically different, but not isolated from each other. Potentially, they will "borrow" words and utterances from each other. In any case, we will investigate and analyse how "heteroglossia" and "the other's words" are used in the different discourses.

To answer the question of how scientific facts are discursively and linguistically constructed in the vaccination debate in printed news and digital media, we combine political and dialogical discourse theory together. This allows us to analyse the mediated construction of truths and facts in the vaccination debate during the Covid-19 lockdown in terms of how hegemonic scientific positions and counter-hegemonic positions following a populist logic develop both discursively and linguistically. We will examine how hegemonic scientific social imaginaries and contesting myths are articulated through the complex interplay of authoritative and internally persuasive discourses in which different voices draw on power, knowledge, and counter-knowledge. We also examine how various media draw on different discursive and linguistic practices in scientific conflicts.

Methods

The study focuses on the articulation of Covid-19 vaccine viewpoints across the hybrid media environment and is based on media data combining a selection of articles from Danish print newspapers with posts and comments from Facebook groups with a focus on discussing the Danish government's Covid-19 policies during the pandemic lockdown period.

The data consist of: (1) All articles published containing the word "vaccine" from the periods 27.12.2020–09.01.2021 (where the vaccination programme was introduced) and 11.03.2021–18.03.2021 (when some of the vaccines were paused) in the Danish broadsheet newspapers *Politiken* and *Berlingske* and the tabloid newspaper *Ekstra Bladet*. We included news articles, debate pieces, and comments. Data from period 1 include 270 articles, while period 2 consists of 145 articles. The three newspapers – all among the largest in Denmark – have been chosen to

Table 7.1 Newspaper articles

Period 1:		*Period 2:*	
Berlingske (Ber)	19 articles	Berlingske	9 articles
Ekstra Bladet (EB)	14 articles	Ekstra Bladet	12 articles
Politiken (Pol)	19 articles	Politiken	13 articles

reflect both the broadsheet and the tabloid press across the political spectrum thus including both elite and popular positions. *Politiken* is a culturally oriented centrist newspaper, *Berlingske* a right-wing business newspaper, while *Ekstra Blandet* is, in its own words, for "the man on the street".

Both the introduction and the withdrawal of the Covid-19 vaccines are interesting periods for this study. They illustrate the promise of science and new technology as a possible solution to the crisis in the first period, while this promise was (partially) compromised during the period of withdrawal of some vaccines, highlighting a scientific uncertainty and political disagreement about the implementation of science in society. From a total of 415 articles, we selected 86 articles for the discourse analysis using the following relevance criteria: To directly address the issue of "vaccination" and to contain discussions and points of view about science and scientific facts.

Table 7.1 gives an overview of the data from newspapers.

(2) The Facebook data drawn on three Danish groups: *Men in Black*, *Folkebevægelsen for Frihed* (The People's Movement for Freedom), and *Kend din Grundlov* (Know Your Constitution). All of these represent groups that take a stance against the official vaccination programme. The first two arose in the wake of the extensive societal shutdown during the pandemic in 2020; the last relates to a broader scepticism towards the Danish Constitution, which has inscribes the Monarch as the signatory of all laws adopted by the Parliament and thus formally attributed power. We analyse four selected posts with the accompanying images, videos, likes, and comments. The comments that are analysed are selected according to the same relevance criteria as the newspapers: To directly address the issue of vaccinatio' and to contain discussions and point of views about science and scientific facts.

Table 7.2 gives an overview of the data from Facebook.

The newspaper articles and the Facebook pages comment threads are analysed using a combination of a linguistic and discursive strategies to understand how the central antagonistic frontiers in the vaccination debate were articulated according to the political and scientific logics. To do this, we conducted a qualitative, conceptual coding of all the textual material based on the main concepts: Scientific social imaginaries, myths as hegemonic knowledge and counter-knowledge, populist scientific logic, dialogism, heteroglossia and centrifugal/centripetal forces in discourse and language. We compare both the different newspapers among themselves

Table 7.2 Facebook posts

Post	*Date*	*Group*	*Comments/shares*	*Reactions*
"It is more and more clear that mass vaccination was the goal from the very beginning..."	28.11 2020	Folkebevægelsen for Frihed (People's Movement for Freedom): FFF www.facebook.com/groups/263242591426591 (accessed 01.12.2022)	143 comments 56 shares	364 likes
"Anders Lund Madsen got a job in a vaccinations centre…"	12.03 2021	Folkebevægelsen for Frihed (People's Movement for Freedom): FFF www.facebook.com/groups/263242591426591 (accessed 18.11.2022)	321 comments 13 shares	515 reactions
"i feel like asking Corona's witnesses if they still believe that we are ridiculous tin foil hats?!"	11.03 2021	Kend din Grundlog (Know Your Constitution): KdG www.facebook.com/groups/3022919607818708 (accessed 17.11.2022)	32 comments 14 shares	53 reactions
"Terrible what is happening – let's hope that no more will suffer"	11.03 2021	Men in Black: MiB www.facebook.com/MenInBlackDK/ (accessed 05.10.2022)	332 comments 119 shares	396 reactions

and their modes in the two periods, as well as differences (and similarities) between the newspapers and Facebook groups. The three Facebook groups will only be related to each other to a limited extent.

Analysis

The hegemony of science: Science as a political discourse of power

Articulations of science constitute a privileged moment in the newspapers' coverage of the launch of the Covid-19 vaccines in December 2020 and during the withdrawal of the AstraZeneca vaccine in March 2021. Despite the overall hegemonic power of science to create political legitimacy, however, the newspaper coverage of science during the two periods studied does not adhere to a unified genre, but rather presents a multi-vocal representation of different voices and genres embedded within the journalistic news genre that encompasses both broadsheet and tabloid newspapers. There are short Reuters articles (mainly in *Ekstra Bladet*) as well as opinion pieces and longer articles. Especially the latter is itself an expression of a polyphonic genre, where different experts are interviewed in conjunction with explicit or implicit references to the authorities' official statements, all held together by an evaluative journalist.

In the first period, when the vaccine was first introduced, the press covers success stories about vulnerable and elderly people receiving their first vaccination and shows how well and quickly the vaccination works and has been rolled out (EB 27.12.21; Ber 28.12.20c; Ber 28,12,20b; EB 28.12.20 c; Pol 28.12.20a; Pol 09.01.21a;). Only *Ekstra Bladet* questions the vaccination programme, namely regarding the commercial actors (EB 05.01.21). The challenge faced by the global community is put into a context as a "historical milestone" (Ber 28.12.20e), where there now seems to be light at the end of the tunnel (Pol 02.01.21a). Amongst the most prominent sources quoted in during the vaccine rollout were front-line medical staff who with the vaccine has received a highly anticipated new defensive weapon in the metaphorical war-battle against the Covid-19 virus. While the Covid-19 virus was "an unknown entity" (Ber 27.12.20a), the vaccines now constitute a (scientific) promise for new hope and safety (28.12.20e). Furthermore, portraits are drawn of the diligently researching vaccine pioneers who have cracked the vaccine code (e.g., Pol 10.01.21b).

However, there is also room for voices of doubt, especially regarding the vaccination of children, even though no uncertainty is voiced about the benefits of vaccination (Pol 27.12.20b). Often, the angle is how to convince the doubters (Ber 04.01.21a). The coverage of Covid-19 sceptics is therefore somewhat sporadic and distant, even if a journalist has been following a critical activist and demonstrator for several months; instead of reporting what the current actor experiences, the article presents psychological and social psychological explanations regarding vaccine sceptics and conspiracy theorists, with the former seen as expressions of the latter (Ber 30.12.20a; EB 04.01.21b). Rather than representing a polyphony of heterogenic voices, the sceptics are represented in a way that unifies the discursive horizon of meaning into a hegemonic social imaginary in which science represents the only legitimate way forward in combatting the virus.

Furthermore, the first period is characterised by the press's coverage of Covid-19 vaccines according to an administrative logic focusing on the rollout of vaccination programme. This includes questions of priority and capacity, whether enough vaccine factories have been built, if politicians acted too slowly 6 months ago (Ber 01.01.21a; EB 04.01.21a), what the expected vaccine effectiveness will be, and, finally, what the risk of mutations entails (Pol 10.01.21b) notably with the first sign of a mutation (EB 03.03.21) and the new restriction hereof (EB 06.01.21). The potential scientific uncertainly of virus mutations, at this time the "British variant" potentially a "superspreader", may challenge the vaccines' promise of safety by emphasising concern voiced by named experts in "biocomplexity", "CEO of BioNtech" and "professors" (Ber 29.12.20). The angle is not only related to physical health and risks but also to economics, emphasising the cost to society of being locked down, as well as to social well-being, especially for young people (Ber 06.01.21b). The issue of vaccines is thus presented according to a hegemonic managerial myth of control (Roslyng, 2011): The countermeasures introduced in the face of the uncontrollable risk posed by the Covid-19 virus and it's possible mutations make up a governance system of controllability presented by the public authorities which is, in turn, re-mediated by the broadsheet media in particular.

The second period, after AstraZeneca vaccines were paused on 11.3.2021, represent a challenge to the legitimacy and the hegemony of science, as well as the myth of controllability of the Covid-19 risk by the vaccines. The most significant and striking feature in the selected newspapers during this time is the problematisation of the rationale behind the authorities' decision to withdraw the vaccine (e.g., Ber 12.03.21; Ber 16.03.21, Pol 17.03.21; EB 12.03.21). All three newspapers raise questions about whether there may be a connection between the Covid-19 vaccination and the few cases of blood clots. They quote experts who argue for this viewpoint, while they also point out that the cases of blood clots among those vaccinated with AstraZeneca vaccines are not proportionally higher than in the general population indicating that the cases of blood clots and deaths caused by them are random. The newspapers moreover point out that the EMA (European Medical Agency) does not recommend stopping the vaccine. The fact that several of the experts still support the withdrawal of the AstraZeneca vaccine due to uncertainty does not receive much attention (e.g., EB 14.03.21).

An example of the representation of scepticism and scientific uncertainty is *Politiken*'s article from 17.3.2021 – a few days after the AstraZeneca vaccine was put on hold, when the change has settled, and the press no longer writes quick assessments. Nevertheless, the article states sharply as a kind of conclusion to its investigation that includes a series of expert interviews:

> AstraZeneca's vaccine can be acquitted of causing more blood clots, but experts are closely examining the suspicious cases of a very unusual blood reaction.
>
> If the law of large numbers was the only truth about vaccines, then all of Europe would continue to inject needles with AstraZeneca fluid into the arms of tens of thousands of citizens every day. If you look at how many cases of infection – and thus deaths and long-term effects – the vaccine prevents, the number of suspicious side effects is drowned out.
>
> (Pol 17.03.21)

The other's word dominates the entire article: AstraZeneca is "acquitted" as if it were a trial, and the quoted experts act as witnesses, drawing on their "foreign" scientific experience and knowledge. Despite the multi-vocality that threatens the legitimacy of science, the scepticism represented above thus becomes an assumed scepticism, in which scientific language draw a new and slightly different horizon of meaning, albeit one that still mainly represent a supportive position in relation to the governmental vaccination programme.

In addition to this positioning, a significant portion of the articles in the second period address the international reaction to the blood clot cases (e.g., EB 12.03.21;

EB 15.03.21; EB 18.03.21; Pol 16.03.21; Pol 17.03.21), as well as the uncertainty it has created among those who have already received an AstraZeneca vaccination (Pol 12.03.21). Finally, *Ekstra Bladet* publishes a critical post about *Men In Black* (EB 15.03.21).

Overall, the press in the first period almost unconditionally supports the authorities and their strategy for the treatment of Covid-19 thus underscoring the myth of scientific and managerial control of the risk of the Covid-19 virus. The newspapers mainly report success stories along with data and material that support the government's actions. People's doubts about vaccination are mentioned as understandable, but they must be overcome. Vaccine resistance, on the other hand, is almost treated as a disease to be eliminated. Specifically, *Berlingske* and *Politiken* speak with an authoritative discourse, influenced by hegemonic and centripetal forces.

This cannot be said of *Ekstra Bladet*, as their jargon is more direct and sharper. Vaccine sceptics are referred to not only by the derogatory term "tin foil hats", which has become their nickname, but also as "conspiracy idiots" (EB 04.01.21b), and in one opinion piece, the change in the vaccination programme is called "crazy" and "stupid" (EB 07.01.21). Here, the voice of rationality does not prevail, but rather communication is done from a position that allows and embeds emotions. The tabloid newspaper is thus set apart from the other mainstream media, and share instead some similarities with online media, as we will see later. In general, throughout the entire first period in all three newspapers, explicit and detailed scientific argumentation and explanation are given little coverage. However, at the same time science and scientists are the heroes, almost mythologised.

The second period is different. Here, the newspapers directly present a counter-discourse to the authorities' hegemonic statements which they predominantly supported previously in period one. They engage with a foreign and more detailed scientific discourse to problematise an apparently authoritative discourse, and they do so without paying attention to the uncertainties present in the scientific discourse. In other words, newspapers want unequivocal statements and truths to support their own opinion and position, and they use science for that purpose.

The analysis thus shows that the media discourse invokes no open investigation of a problem, but the topic is politicised, narrow-minded, and monological – under the "law of large numbers", just as the argument for putting the vaccine on hold is not heard. An article explains that it is only under special circumstances that a specific type of side effect occurs, "an unusual condition" where the blood clumps together in clots that use up the platelets. The few people in this risk group "drowns," as stated in the initial summary, and could easily be excluded from AstraZeneca's vaccines. The head of the European Medicines Agency, Emer Cooke, is quoted for saying: "The benefits of the vaccine outweigh the potential risk" (Pol 17.03.21). In mediated discourses, the result is an objectification of science and scientific statements, and the dialogic plurality and diversity within the scientific discourse itself disappear. For the newspapers, the authorities represent an anti-progressive discourse of caution. That is, the alternatives are a monologic discourse based on "scientific evidence". Or rather: In the press, science is expressed as an authoritative discourse that can reinforce the expressive power of the political discourse.

Counter-knowledge as "the people's" knowledge

The three Facebook groups selected for this study are all focused on providing a platform for like-minded people to express their concerns about the Danish government's anti-Covid-19 policy measures. All three groups share an emphasis on the resistance to the proposed vaccination programmes. In the following, we look at how the anti-vaccination stances presented are part of a discursive, political construction of counter-knowledge that focuses on "the people's" science drawing on certain populist logics. Meanwhile, this articulation of counter-knowledge also presents its own scientific social imaginaries that draw on particular representations of facts.

Initially, it must be pointed out that it is characteristic, but not surprising, that the communication on the three Facebook pages is idiosyncratically oriented. This means, for example, that when one person makes a value judgement, it is confirmed or repeated in varying forms by another person, who then seeks confirmation from the first person, after which a third person makes a new parallel value judgement without any reference at all to the earlier exchange, which is then confirmed by a fourth person. Only occasionally are there discussions that involve multiple exchanges or more than two to three people at the specific exchange. It is also rare for a statement to contain a longer argument. However, longer passages are often quoted from experts who are sympathised with, or from laypeople who present more elaborate arguments or have created opinion pieces, video material, or similar. Typically, therefore, is therefore no dialogical exchange and testing of, or building of, arguments.

An example of the above is the long thread on the *People's Movement for Freedom* (FFF 21.03.21), which reacts to the comedian and TV host Anders Lund Madsen, who is known for his anti-authoritarian attitude, volunteering to publicly get vaccinated with AstraZeneca vaccines just before it is withdrawn; in that context, he advocates getting vaccinated and as the icing on the cake, he has been hired in a vaccination centre. The reactions are full of outrage with predicates like "disgusting", "tasteless", "fool", "ridiculous idiot", but also (even) more aggressive statements like "fool – fucking traitor", "the evil person", "free us from evil and more brainwashing", and "dirty, cynical and manipulative propaganda". However, several comments target the act of getting vaccinated and the vaccine itself more broadly: "the dictatorial red criminal syndicate" and comparisons with Goebbels and Himmler and Nazi Germany in general: "the worst atrocity committed against humanity since World War II" and "vaccine macht frei" ("vaccine provides freedom"). While the thread only minimally discusses ideas and perspectives, it forms a monolithic unit of viewpoints, or rather, a singular viewpoint. The same position is repeated again and again, only varied to a lesser extent, often in terms of style and degree of aggression.

A consensus is expressed in the three Facebook groups by the fact that the space for expressive freedom is narrow, and thus what is deemed acceptable is very limited. This becomes particularly evident when someone uses a metaphorical or ironic form of expression, which can be seen in the following exchange from the *People's Movement for Freedom* (FFF 12.03.21):

XX: One can also say that the vaccine separates the sheep from the goats, in the sense that those who are so stupid to take the vaccine die, and those who are smart enough not to, survive. Then there are only intelligent, critical-thinking people left. I wonder how the elite will handle that? 👍 11

YY: XX Congratulations, young man.
You have figured out our plan!!

XX: YY "Your" plan? What the hell are you talking about? You have nothing to do with what is happening now, and you are not part of the elite! You are a fucking loser with delusions of grandeur suffering from narcissistic personality disorder! Fuck, you're a disgusting human being! 😫😫😫

YY: XX Calm down! It was meant to be ironic.

XX: YY Are you sure about that?

It is striking how aggressively XX approaches YY. XX has no trust in YY's ironic intention but takes the statement literally and cannot relate to a non-literal expression. This is one among other illustrative examples of how limited the communicative space is. Thus, the language used cannot be characterised as a creative or a decentralised living language, but paradoxically, as a unit language within an authoritative discourse.

Another way in which the narrow communicative space is expressed is through the categories that are set up. Across all the posts analysed, "the people" becomes a empty signifier positioning themselves against the political system, for example: "Us Danes are cleverer than Mette Frederiksen had figured out". And in reply: "Exactly! The woman is definitely wrong! Maybe, she after all, lacks experience. From real life! (Laughing smiley)" (FFF 28.11.20). "The people" constructed here is, on one level, a national entity, as the group members express their scepticism on behalf of the whole population. Simultaneously, "the people" represents a specific group brought together in a political community by their shared opposition to the Covid-19 measures and their ability to see through government policies as discussed on the *Men in Black* Facebook group: "Most people have not realised that they are lab rats and there will be no compensation when it goes wrong" (MIB 11.03.21). Several members position themselves in relation to the mainstream public by reclaiming the label 'tin foil hats': "Yes, "funny thing", the "tin foil hat people" have until now been correct about all their warnings. You cannot trust a government blindly that is becoming more and more dictatorial". Furthermore, "Tin foil hats will have something to be proud of when the history books are written", while another writes proudly: "I am a tin foil hat! Today a tin foil hat is an enlightened person… Someone who spends time trying to find head and tail in the whole circus" (MIB 11.03.21). Both the shaming of the Prime Minister Mette Frederiksen, the government, and its sympathisers with a series of insults, as well as the conspiracy theoretical suggestions regarding what may be in the vaccines, serve as a centripetal force in creating and continuously affirming a sense of community. Here, there is a united front against the hegemonic power and its supporters.

This political and ideological counter-knowledge position should not be viewed as an anti-science discourse. In fact, there is a strong reliance on scientific argumentation across many comments. The articulation of science is central to all three Facebook groups in their discussions on the Covid-19 action programme but differs to a limited extent from the dominant pattern outlined in the previous text: Here we find longer statements, primarily quoted text from other sources, such as expert statements or texts by experts; that is links to other media sources (often but not always well-renowned mainstream media) to support vaccine scepticism in which scientists express concerns about possible side effects of vaccines are quoted. This form of intermediality lends legitimacy to the facts presented on Facebook, for instance, in a comment with a link to an interview with a scientist from the Danish broadcast TV station.

Quoted material rarely leads to responses beyond likes, as seen in the post: STOP THE VACCINE IMMEDIATELY! (FFF 12.03.21). It consists of a translation of the virologist Geert Vanden Bossche's appeal to "all authorities, scientists, and experts worldwide" regarding the risk that the vaccines do not cover Covid-19 mutations, but rather stimulate a chain of mutations that would then be impossible to contain. – Although Bossche himself claims not to be opposed to vaccines as such, it still represents a strong criticism of vaccination programmes worldwide, as he warns that "prophylactic" (i.e., disease-preventive) vaccines used in mass campaigns during viral pandemics will lead to resistance to mutations, known as viral escapes. In other words, it represents a scientific argument against the current vaccines, although the actual scientific basis in the argument is absent but relies on citing scientific evidence to support the claim. Nevertheless, the post stands out significantly from the previously mentioned posts, due to its length and the use of technical terms (in addition to prophylactic and viral escapes, also e.g., spike protein and mRNA function). Therefore, it is a text characterised by a foreign discourse and a (seemingly) developed argument. The text delves into the realm of hegemonic discourse and establishes a high ethos through its technical terms and by frequently highlighting Bossche's academic credentials as well as several links. While the text does adopt a professional and serious tone at points, it is also in this aspect that the text as a counter-discourse is filled with emotive charges: "The VACCINE will create a wild pandemic and cause a global catastrophe", "scientists and clinicians are blinded", the developed vaccines are "completely inappropriate" etc. (FFF 12.03.21). One might be tempted to describe it as an internally persuasive discourse, as an expression of a centrifugal force, as personal distress combined emotive charges functions as foreign words within a scientific discourse, thus erupting as a call for help to "all authorities and scientists." However, with its pathetic language ("I assure you that..." etc.), the text predominantly centers around Geert Vanden Bossche, the author himself: emphasising that he does not typically utilise platforms to discuss vaccine-related issues, that he is a dedicated virologist and vaccine expert, and that he has presented his "findings" at a Vaccine Summit and to the WHO, etc. The text thus serves as an illustration of mixed language – aiming to instruct, persuade, warn, and promote the author himself. In this manner, it displays heteroglossia, but paradoxically appears monologic. Hence, the instructive and invoked voices do not create a struggling chorus of diverse voices, but rather converge towards a direction as a centripetal force of language.

It is highlighted that the criticised science is the hegemonic science confirmed by the authorities. In a post in the *People's Movement for Freedom* Facebook group from November 2020, as the vaccine programme was being gearing up for implementation, the scepticism towards the science presented by the Danish government unfolds as follows:

> It is more and more clear that mass vaccination was the goal from the outset. The issure then becomes: How to sustain a heightened sense of fear and artificially inflated infection rates to the extent that the populace demands a vaccine and consents to voluntary vaccination. This is achieved by obstructing access to genuine information and impeding independent experts' evaluation of the Covid19 threat.
>
> (FFF 28.11.20)

The post sparks a large number of supportive comments in which a major theme in the critique of the Covid-19 vaccines pertains to the uncertainty as to what the vaccines contain. "JUST WAIT AND SEE, I AM SURE THAT THERE IS PROBABLY MORE THAN VACCINE THAT WILL BE INJECTED" (FFF 28.11.20). Another user reads between the lines and questions how feasible it is to produce waterproof microchips that go through an injection needle. He points out that he is not pro-vaccine but genuinely interested in knowing the answer. His query leads to suggestions such as nanotechnology and hypergel. A link is also presented to research from John Hopkins that has designed a special microdevice that can deliver drugs into the body. Further down in the comment thread someone asks directly what the vaccine contains and receives multiple answers: "aluminium", "virus from a chimpanzee that will change our DNA", "or is it from Mink?", or "you will never know, it is top secret, doesn't that say it all" (FFF 28.11.20). It is significant that there a strong united front against the authorities and vaccines, while there are also a diverse range of suggestions about what lies "behind" the vaccines, most notably what is *in* the vaccines. We see here the mentioned antagonistic frontier (Laclau, 1990) between, on the one side, the scientific foundation of the vaccination programme as promoted by elite actors such as the government and their related institutions (researchers, media, and international networks), and, on the other side, an alternative and populist form of science based on counter-knowledge that gains its legitimacy from the people.

The withdrawal of AstraZeneca from the national vaccination programme due to possible serious side effects such as blood clots further emphasised the antagonism between an (illegitimate) elite science and a (legitimate and sensible) people's counter-knowledge as seen in a comment in the *Know Your Constitution* Facebook group. Linking to an article from the internet-based media dagens.dk, the comment states: "I feel like asking Corona's witnesses if they still think that we are ridiculous tin foil hats" (KdG 11.03.21). The withdrawal is presented as a vindication for the counter-knowledge position and accompanying the comments show clear support: The fact that the government proposes voluntary fast-tracking of vaccinations for individuals who do not mind the possible risk of side effects is seen as a "biological roulette" that is moreover rejected by private vaccination clinics: "Finally some doctors with sense who contradict our government!!!" And the comments concluded that: "This is an assault on the population". And a side-effect presented as a statistical fact is stated as: "sterility in 97% women" (KdG 11.03.21).

Overall, the communication on three Facebook groups is notable for a striking lack of diversity; different genres are drawn on, experts from different platforms are quoted, but in a way where all statements point in the same direction. An immediate heteroglossia appears as a monologism; a counter-discourse that could have manifested a creative inner persuasive discourse turns out to be a unitary language that regulates communication with a centralising, unifying and assimilating effect. The presentation of counter-knowledge statements on Facebook form a counter-hegemonic discourse that is political and is characterised by communicating in a number of clarified categories – the people versus the elite, the truth-seekers versus the liars and the cover-ups, etc. Here, the people's science becomes an essential piece that functions as an empty signifier. All in all, this draws a populist logic.

Conclusions: Knowledge conflicts in hybrid media

Based on the analysis, we argue for two main characteristics in the way both mainstream media and critical Facebook groups use scientific positioning: In news media, it is significant that authoritative elite discourses about vaccination draw on a scientific social imaginary, which establishes truths in line with the government's goals. This is what we uncovered in the initial period. Equally striking, however, is that mainstream media are also the first to be critical of authorities when, based on scientific uncertainty, they become cautious and pause some of the vaccines. Here, scientific argumentation functions as critical evidence for an unambiguous approach to the vaccination issue – as exemplified by the authorities' use of risk-benefit analysis. Mainstream media are thus more centripetal and monological-unitary than the authorities. Both have the same goals, but in the media, the scientific social imaginary serves to make discussions unambiguous and dismiss doubt, thus attempting to be more authoritative than even the hegemonic discourse from the government and authorities.

In the Facebook groups studied, it is evident how online anti-vaccination groups use a counter-language that appears to draw on a liberating and creative counter-discourse. This counter-discourse creates new connections and interprets data in radical and sometimes surprising ways. However, this form of counter-knowledge is still dominated by hegemonic and authoritative ideological discourses, showing little responsiveness to opposing arguments. The potential for considering diverse representations and arguments is almost excluded. What could have been an alternative space for diverse viewpoints, a creative and internally persuasive discourse with centrifugal forces and a diverse heteroglossia, turns out to be very limited, more monological, and more centripetal than both the authorities and the mainstream media represent. An important aspect of the Facebook groups is the continual reaffirmation of belonging to a political alternative knowledge position engaged in a populist form of social media activism that divides the social space into elite versus people's science. Additionally, science is used for this purpose, as the group presents itself as capable of discerning what "true" people's science says and thus reinforces the myth of the scientific social imaginary.

It is striking that both mainstream media and Facebook groups use science in an authoritative manner to dismiss doubt and discussions and insist on monologue: only one truth. The scientific discourse, for its part, is open to counter-arguments, but this mode is weakened when it also becomes the word of the authorities. It is thus paradoxical that scientific discourse, even in public, appears most credible when it can be held in the uncertainty that is its premise.

References

Agustín, O. & Nissen, A. (2022). The anti-restriction movement and the populist counterpublics in Denmark. *Partecipazione e Conflitto, 15*(3), 71–741. https://doi.org/10.1285/i20356609v15i3p741

Bakhtin, M.M. (1981). *The dialogic imagination.* University of Texas Press.

Bakhtin, M.M. (1984). *Problems of Dostoevsky's poetics.* University of Minnesota Press.

Bakhtin, M.M. (1986). From notes made in 1970-71. In *Speech genres and other late essays.* University of Texas Press.

Bolisani, E., Navarro, J.G.C. & Garcia-Perez, A. (2021). Managing counter-knowledge in the context of a pandemic: challenges for scientific institutions and policy-makers. *Knowledge Management Research & Practice 19*(4), 517–524. https://doi.org/10.1080/14778238.2021.1911606

Bridgman, A., Merkley, E., Loewen, P., Owen, T, Ruths, D. Teichmann, L., & Zhilin, O, (2020). The causes and consequences of COVID-19 misperceptions: Understanding the role of news and social media. *Harward Kennedy School Misinforation Review*, June 18. https://doi.org/10.37016/mr-2020-028

Cegarra-Navarro, J.-G., Bolisani, E. & Cepeda-Carrion, G. (2022). Linking good counter-knowledge with bad counter knowledge: the impact of evasive knowledge hiding and defensive reasoning. *Journal of Knowdedge Management, 26*(8), 1367–3270. https://doi.org/10.1108/JKM-05-2021-0395

De Cleen, B. (2018). Populism, exclusion, post-truth. Some conceptual caveats. *International Journal of Health Policy and Management, 7*(3), 268–271. https://doi.org/10.15171/ijhpm.2017.80

Demata, M. (2025). Fighting (for) truth? Alex Jones, the WHO and the legitimazation of consipiracy discourse. In M.M. Roslyng, A. Rantasila, & A.M. Jönsson (Eds.), *Communicating science, climate change and the environment in hybrid media: Constructed facts, contested truths* (pp. 91–111). Routledge.

Eslen-Ziya, H. & Giorgi, A. (Eds.) (2022). *Populism and science in Europe.* Palgrave Macmillan.

Eslen-Ziya, H., McGarry, A., Jenzen, O., Erhart, I. & Korkut, U. (2019). From anger to solidarity: the emotional echo-chamber of Gezi park protesters. *Emotion, Space and Society, 33*, 100632. https://doi.org/10.1016/j.emospa.2019.100632

Goldenberg, M.J. (2021). *Vaccine hesitancy: public trust, expertive, and the war on science.* University of Pittsburgh.

Hart, P.S., Chinn, S., & Soroka, S. (2020). Politicization and polarization in COVID-19 news coverage. *Science Communication. 42*(5), https://doi.org/10.1177/1075547020950735

Jalaludin, F.W., Loh, Y.X., Seah, C.S., Chang, Y.K., Muhammad, E.E. & Chin, W.Y. (2024). Individual response to the spread of misinformation about COVID-19 via Facebook and WhatsApp. *Procedia Computer Science, 234*, 962–969. https://doi.org/10.1016/j.procs.2024.03.085

Laclau, E. (1990). *New reflections on the revolution of our time.* Verso.

Laclau. E. (2005). *On populist reason*. Verso.

Lelisho, M.E., Pandey, D. Alemu, B.D., Pandey, B.K. & Tareke, S.A. (2023). The negative impact of social media during COVID-19 pandemic. *Trends in Psychology, 31*, 123–142. https://doi.org/10.1007/s43076-022-00192-5

Lewandowsky, S., Ecker, U.K.H., & Cook, J. (2017). Beyond misinformation: understanding and coping with the "Post-Truth" era. *Journal of Applied Research in Memory and Cognition, 6*(4), 353–369. https://doi.org/10.1016/j.jarmac.2017.07.008

Madvig, F., Achiam, M., Adler-Nissen, R., Johansen, N., & Whiteley, L. (2022). Coming closer to citizens? frustrated dialogue on the Danish health authority's Facebook page during COVID-19. *Science and Environmental Communication, a section of the journal Frontiers in Communication, 7*, 822471. https://doi.org/10.3389/fcomm.2022.822471

Makinen, M. (2025). Citizen activists or pandemic deniers? Alternative voices in the Finnish journalistic media during the COVID-19 pandemic. In M.M. Roslyng, & A. Rantasila– (Eds.), *Communicating science, climate change and the environment in hybrid media: Constructed facts, contested truths* (pp. 131–149). Routledge.

March, K.J., Reyes, R.S., Pentz, B., et al. (2021). News media coverage of COVID-19 public health and policy information. *Humanties & Social Sciences Communications, 8*(1), article 220. https://doi.org/10.1057/s41599-021-00900-z

Mouffe, C. (2005). *On the political.* Routledge.

Niemiec, E. (2020). Covid-19 and misinformation. Is censorship of social media a remedy to the spread of medical misinformation. *Science & Society. EMBO reports, 21*(1–4), e51420.

Pool, J., Fatehi, F., & Akhlaghpour, S. (2021). Infodemic, misinformation and disinformation in pandemics: scientific landscape and the road ahead for public health informatics research. *Public Health and Informatics.* European Federation for Medical Informatics (EFMI) and IOS Press. https://doi.org/10.3233/SHTI210278

Raffini, L., & Pealva-Verdú, C. (2022). The problematic relationship between science, politics and public opinion in late modernity: the case of anti-vax movement in Spain and Italy. In H. Eslen-Ziya & A. Giorgi (Eds.), *Populism and Science in Europe* (pp. 141–162). Palgrave Macmillan.

Roslyng, M.M. (2011). Challenging the hegemonic food discourse in Britain. The British media debate on salmonella in eggs. *Science as Culture, 20*(2), 157–182. https://doi.org/10.1080/09505431.2011.563574

Roslyng, M.M. (2022). Between populism and popular citizenship in science conflicts. In H. Eslen-Ziya, & A. Giorgi (Eds.) *Populism and Science in Europe* (pp.207–229). Palgrave Macmillan.

Roslyng, M.M., & Larsen, G. (2021). The struggle between the power of language and the language of power. Pro- and anti-vaccination memes and the discursive construction of knowledge. *Journal of Language and Politics, 20*(6), 894–912. https://doi.org/10.1075/jlp.19101.ros

Schmidt, H. (2023). Pandemics and politics. Analyzing the politicization and polarization of pandemic-related reporting. *Newspaper Research Journal, 44*(1), 26–53. https://doi.org/10.1177/07395329221095850

Sismondo, S. (2017). Post truth? *Social Studies of Science, 47*(1), 3–6. https://doi.org/10.1177/0306312717692076

Speed, E., & Mannion, R. (2017). The rise of post-truth populism in pluralist liberal democracies: challenges for health policy. *International Journal of Health Policy and Management, 6*(5), 249–251. https://doi.org/10.15171/ijhpm.2017.19

Thompson, D. 2008. *Counter-knowledge.* Atlantic Books.

Vraga, E.K. & Bode, L. (2020). Defining misinformation and understanding its bounded nature: using expertise and evidence for describing misinformation. *Political Communication, 37*(1), 136–144. https://doi.org/10.1080/10584609.2020.1716500

Ylä-Anttila, T. 2018. Populist knowledge: 'Post-truth' repertoires of contesting epistemic authorities. *European Journal of Culture and Political Sociology, 5*(4), 356–388. https://doi.org/10.1080/23254823.2017.1414620

Zimmerman, S., Gaugler, J.E., & Nkimbeng, M. (2021). COVID-19 vaccination and implementation science: how one can benefit the other. *Journal of the American Medical Directors Association, 22*(11), 2223–2224. https://doi.org/10.1016/j.jamda.2021.09.018

8 Citizen activists or pandemic deniers?

Alternative voices in the Finnish journalistic media during the COVID-19 pandemic

Maarit Mäkinen

The time of the coronavirus pandemic has been a special challenge for politics, democracy and the media. Policy makers have been faced with quick decisions, and the media has tried to keep up. During the pandemic, Finns have spent more time with different media sources. (Kohvakka & Saarenmaa, 2021). According to a public survey, media has a lot of power in the Finnish society, and most people think that this power has increased recently. At the same time, evaluation of the social performance of mainstream journalism have become more critical. "Evaluation of the success of media in promoting Finnish culture and cohesion has weakened the most. Trust in the independence of media and its ability to monitor the exercise of power is also weaker than before" (Matikainen et al., 2020 p. 5).

During the corona pandemic, freedom of speech was restricted in many countries in order to prevent the spread of disinformation. Social media channels that clearly deviated from authority communication or criticised it, or discussed, for example, the safety of coronavirus vaccines, were warned or shut down also in Finland. According to studies on the exceptional period, Finns have had a relatively high level of trust in the news media, but there has been criticism about the inconsistency of communication and the lack of attention paid to the opinions of citizens (Tilastokeskus, 2022; Hakala & Ruggiero, 2022; Jallinoja & Väliverronen, 2021; Simonen et al., 2021). The citizens' trust in the institutions of society has somewhat declined (Metelinen, 2021). According to the OECD Observatory of Civic Space report, the Finn's trust in different institutions varies, and only a small portion of people believe that they can affect political processes (OECD, 2021). The report recommends equal consideration of opinions and a broader dialogue with citizens.

Political and ideological polarisation has recently been a topic of interest for researchers in Finland and internationally (Fletcher et al., 2020; Reunanen, 2022). The exceptional period has also raised concerns and debate in many countries about the polarisation of communication, which refers to the juxtaposition and tensions between people (Tolonen, 2021; Niemi, 2022; Reiter-Haas et al., 2022). Polarisation and differing views are a part of all societies, but when they become too strong, they can cause problems. If groups of people are pitted against each other on the basis of supposed characteristics and the setting is seen as black and white, the situation can cause social conflicts. There is an "us and them" mindset,

DOI: 10.4324/9781003479550-10

which weakens the possibilities of cooperation in crisis situations and threatens social harmony (Brandsma, 2017, p. 104; Attias & Kangasoja, 2021). The exceptional period has also been described in public discussion as a social pandemic that breaks down relationships between people and excludes groups of people from society.

Polarisation of communication has been visible, for example, in heated debates about the benefits or harms of coronavirus vaccines and restrictive measures. Polarisation in communication has been identified in particular on social media platforms and discussion groups, which has led to the removal of voices that are critical of administrative communication (Niemiec, 2020; Farr & Rodriguez, 2020). According to scholars examining trust towards the media, the ideological divide in society has become a problem for journalism that considers itself neutral in situations where it has excluded alternative viewpoints (Ojala, 2021; Reunanen, 2022). News reports have emerged in which, for example, expert discourse supporting political decision-making has been juxtaposed with those who are sceptical about coronavirus vaccines, although scepticism about coronavirus vaccines is usually not a matter of being anti-science or even anti-vaccine. (Goldenberg, 2021; Nurmi & Vuolanto, 2020; Väliverronen & Jallinoja, 2021).

The journalistic mainstream media has been criticised for giving too much and one-sided support for authority communication and for excluding people expressing differing opinions. (Mallon & Saalo, 2021, p. 133–145; Siippainen, 2021; Mattila, 2022). Many communication specialists who have assessed the exceptional communication during the coronavirus pandemic have emphasised the importance of participatory and multi-voiced interaction, which has not been fully realised (Kantanen & Koskela 2022). "The pandemic of the unvaccinated" became a phrase that was repeatedly used in Finland and internationally during the exceptional period. The phrase was criticised in the medical journal *Lancet* for being simplistic and for undermining the unity of citizens (Kampf, 2021). The emotional dynamics of polarisation may have been reinforced by metaphors describing the situation as a "war against the virus" (Forsberg, 2020).

At the same time, the exceptional period has generated a large civil movement that is sceptical or critical of authority communication on the pandemic management. This has taken the form of alternative media publications, networking, protests, events as well as various statements. As a result of restrictive measures concerning publishing on social media, critical views have sought out publishing platforms that do not restrict publishing. Researchers have found that the current media environment poses challenges to public debate and to the communication rights of citizens (Reunanen & Herkman, 2020). Challenges are posed by both the revenue model of media companies and the political aspirations of different actors, which are not easy to resolve from a legal nor a freedom of speech perspective.

This article examines whether the Finnish mass media during the exceptional period shows signs of excluding critical views of citizens and of communication that reinforces polarisation. Mass media refers to mainstream media, which is commonly used to refer to traditional news media and high-circulation media that often influence dominant views. The research poses a question: how are those with views that differ from those of the mainstream media portrayed and what kinds of images are constructed of alternative views? Finally, the research considers what kinds of means could be used to achieve a multi-voiced approach that respects different views in media coverage.

Polarisation and the role of mainstream media

According to philosopher Bart Brandsma (2017), polarisation or juxtaposition is a way of thinking that is neutral in itself, but becomes a problem if it becomes black and white and absolute. Polarisation of communication develops from an "us versus them" approach, where groups are pitted against each other on the basis of supposed characteristics. The fuel for polarisation is identity speech, where an identity of a group is defined by speech. In this case, individuals are characterised into groups based on stereotypical characteristics (Brandsma, 2017, p. 26–43). Polarisation is fuelled by stigmatisation and hate speech. Brandsma points out that polarisation works on an emotional level. Researchers often talk about affective polarisation (e.g. Tolonen, 2021). Polarisation is not about facts but about how things feel. Facts can act as fuel if they are interpreted only from one's own perspective while reinforcing one's own thought pattern. Instinctive emotional dynamics cause a defensive reaction, which makes debate difficult and arguing about facts pointless. (Brandsma, 2017, p. 23; Attias & Brandsma, 2021, p. 25.)

According to Brandsma's model, people take different roles in a situation where there are polarised views. The pushers feel like they are morally in the right, and they want their cause to be visible. The joiners seek to join a group, but they do not hold such strict opinions. The silents, on the other hand, see the complexity of issues, even if they do not express their views. (Brandsma, 2017, 32; Attias & Kajanoja, 2021.) According to Brandsma, in the midst of polarisation, the silent ones are the most important in resolving conflicts because they do not add fuel to the fire on either side. Some citizens may remain silent to avoid being stigmatised or socially excluded for their views.

Silence can also signify silent resistance, where polarisation is avoided, but that can manifest itself as avoidance of issues or as hidden disobedience, "nagging in everyday life" (Lehtola & Autti, 2019, p. 8–24). If mainstream media only conveys the presumed majority view, a potentially significant part of opinions will be buried from view and the silent will become increasingly silent. Political scientist Noelle-Neumann describes this trend as "a spiral of silence" (Noelle-Neumann & Petersen, 2004), which at its worst leads to democratic debate withering away. According to van Dijk, a critical discourse analysis scholar, the black-and-white simplification presented in public often carries over to grassroots' level debates, which can lead to a communicative environment that discriminates between citizens (van Dijk, 2008, p. 65–85) and a situation where people only express views that support the consensus.

Attias and Brandsma (2021) describe how unresolved issues can linger, make one's thoughts black and white and tighten tensions. Stigmatisation and silencing of alternative views may drive people into their own camps, limiting the debate to like-minded groups and deepening the polarisation of communication. Conversely, dialogue, openness and acceptance of differences may defuse tensions in communication. (Attias & Brandsma, 2021, p. 45, 104–128.)

What then is the role of mainstream journalism in a tense communication landscape? In most cases, the role lies somewhere between business and social action. Traditionally, the social tasks of the journalistic media have been defined as monitoring the exercise of power and maintaining debate that is a part of democracy. According to research concerning the role of Finnish media in networks of power, media's power is often based on a moral consensus, with little ability to actively address questions where there are genuine political disagreements and broad conflicts of interest. Instead, those in power seek, and in some ways succeed, according to the research, "to subdue the resources of symbolic power of the media and journalism to serve their own efforts". (Kunelius et al., 2010, 461.)

All professional journalists are subject to ethical guidelines which are overseen by the Council for Mass Media in Finland. The task of the council is to interpret good journalistic practice and to defend freedom of speech and publication. According to the council, good journalistic practice is based on the right of everyone to receive information and opinions (CMM, 2022). The member newspapers of News Media Finland are committed to comply with the journalistic guidelines and the principles of the Council for Mass Media in Finland. The tasks of public service media are defined by law, and in addition, their activity is governed by regulations, policies and guidelines. According to the law, "the public service programming shall in particular support democracy and everyone's opportunity to participate by providing a wide variety of information, opinions and debates as well as opportunities to interact" (Act on the Finnish Broadcasting Company, 1993).

Material and methods

This research article examines examples of journalistic news media content in the context of the COVID-19 pandemic, where the polarisation of communication has been particularly prominent between those who trust the authorities' pandemic communication and those who are hesitant about it. The material was collected by following and reading publications in the Finnish mainstream media over the period of 1.1.2021–31.12.2021, when the polarisation of communication was clearly visible in the public. The material was limited to the media content of *Yleisradio* (Finnish Broadcasting Company), *Helsingin Sanomat, Aamulehti* and *Ilta-Sanomat.* The selected news media are members of News Media Finland (a trade association of Finnish news media), and they are committed to comply with the journalistic guidelines and the principles of the Council for Mass Media in Finland (News Media Finland, 2022). Yleisradio is a national media company owned by the Finnish state, and it is committed both to its tasks governed by the law and to ethical principles (YLE, 2022). The monitored content reflected and interpreted the views of citizens who were hesitant about the pandemic management and communication of authorities. In addition to monitoring, archive search tools (e.g. key words "corona", "critical", "activist", "alternatives", "conspiracy"), were used to list a total of about 70 articles or programmes.

Media content presenting alternative, sceptical or critical views of citizens on the pandemic management was selected for analysis. The material thus reflects both the role of journalism in promoting democratic debate and in the polarisation of communication. Monitoring and searching media archives did not reveal any contents where the voices of citizens hesitant about the pandemic communication of authorities were interpreted without correction of the views or polarisation.

The analysis included the television programme *Pandemiankieltäjät (Pandemic Deniers)* produced by Yleisradio as well as 14 newspaper articles that examined the hesitant or critical attitudes towards restriction measures and especially coronavirus vaccines. The articles do not include columns. *Pandemiankieltäjät*, which is a part of Yleisradio's Spotlight series, portrays citizens who are hesitant about the pandemic communication of authorities. The programme was chosen because it was one of the first mainstream media contents to portray citizens who are critical of the management of the pandemic, and because the programme generated a lot of public debate in public columns and especially in social media (e.g. Pitkänen, 2021; Kemppe, 2021). Selected articles from news media also describe views that differ from the communication of authorities and create images of alternative views. The articles act as parallel data, which allows the comparison of recurrences in findings. The selected articles represent examples of content, and similar features were found in articles on the search list that were not included.

In the analysis, the role of media is examined through Brandsma's (2017) polarisation model, which implies polarisation based on the structure of thoughts. In the model, parties with opposing views can be seen as pushers if they feel they are morally right and seek visibility (Brandsma, 2017, p. 26–27). Pushers add fuel to polarisation by stigmatising others and with language use that expresses anger. Emotional dynamics make it difficult for opposing parties to debate, and facts do not resolve disputes. The analysis examines rhetoric expressions and speaker categories of media discourse as well as the choice of different positions.

Media discourse produced by views that differ from those of authority communication and the critical citizens that express them will be analysed from the perspective of framing. Frame analysis seeks answers to the questions of how people with differing views are portrayed and what kind of images of alternative views are constructed. According to communications scholar Erkki Karvonen (2000), the appropriate framework can consciously or unconsciously be chosen for an issue, and thus make it appear as one wants it to. The interpretative frameworks of media are models that are part of journalists' routine work and they enable quick processing, categorising and packaging of information to the public. The task of research is then to ask which framework the story is using and why. (Karvonen, 2000, p. 80.)

All narrative elements, such as images and word choices, form interpretative frameworks. In addition, the media chooses topics for agenda, highlights issues that they consider important and ignores other ones. The prioritisation of issues (agenda setting) influences citizens' understanding of different phenomena and the debate in society.

According to Robert Entman (1993), who studied political communication, journalists' aim for objectivity often leads to reproducing the prevailing interpretative framework, which prevents the public from forming a balanced view of the issue. Framework analysis shows that the meaning that is given priority largely determines the public's interpretation of the issue. According to Entman, political elites are able to control the ways in which societal questions are framed. Appropriate framing creates a desired definition of the situation, which is often reproduced by the media, and in that way it is also possible to steer the public opinion in the desired direction (Entman, 1993, p. 51–58). At the individual level, the effect can be, for example, a change of opinion due to exposure to frames. At the societal level, framing may affect decision-making by shaping public opinion (Ikäheimo, 2021).

The analysis in this research identifies and describes the prevailing frames that portray the views that differ from communication of authorities or question it and the citizens that express them in the mainstream media. A deductive approach has been chosen for the research, whereby the analysed frames are defined on the basis of a preliminary reading of the data. The identification of frames is systematised by applying the list of communication scholar James Willian Tankard (2001, p. 100), which first identifies the possible frames, secondly, lists them, thirdly, develops key concepts and symbols to identify frames, and fourthly, uses frames as categories for content analysis. It is then possible to code media content into these categories. The possible frames have been selected by going through the material and listing frames that can be distinguished and that are commonly recognisable (Tankard, 2001; Ikäheimo, 2021).

The results of the framework analysis are linked to Brandsma's (2017) polarisation model, which examines the relationship between journalistic argumentation and the reinforcement and deconstruction of polarisation. The aim of the description of the frames that are used is to make visible possible unconscious frames in the processes of journalistic work and understanding their impact in the context of an exceptional period.

Activists in a frame of danger

The *Pandemiankieltäjät* programme, produced by Yleisradio, portrays people who are sceptical of the authorities' communication and restrictions concerning the coronavirus (see Figure 8.1). The programme mainly consists of video footage filmed at a protest in Kansalaistori in Helsinki in the spring of 2020, interviews with two people who participated in the protest, expert testimonies from three social scientists and one publically well-known doctor, as well as a journalistic study by the two journalists that made the documentary.

The narrative of the documentary can be compared to film narration, where the viewer is steered into the events and can experience them through narrative

Figure 8.1 The *Pandemiankieltäjät* TV documentary shows a protest in Kansalaistori in Helsinki in March of 2021. In the image, police officers have lined up opposite the protesters. Screen shot from Finnish Broadcasting Company's streaming service *Yle Areena*.

descriptions. When the narrator of the documentary only describes things "as they happened", as in the protest in the documentary, the role of the narrator is constructed in a different way than in direct argumentation. The viewer makes an interpretation within the interpretative framework constructed in the narrative, and the narrator is not accountable for their claim. The narrative guides the viewer to make the categorisation themselves, and the narrator appears as a person who sticks to the facts (Jokinen, 2002, p. 144–145).

The beginning of the *Pandemiankieltäjät* documentary includes descriptions of protesters carrying signs and Finnish flags in sleet, people hugging, police officers in a staircase and escorting a man. Symbolism, such as police officers, flags and banderols activate interpretative frames of danger and conflict in the viewer. In journalistic media, the attention is often focused on conflict, the opponents or the opposition in a given situation (Smith, 1997). The dramatic tension of the programme is created by the choices of shots, music and short interviews, which create an image to the viewer of the activist groups that are portrayed. The impression is emotional and possibly frightening.

The objectives and summary of the *Pandemiankieltäjät* TV documentary are explained in the ingress of the recorded programme:

> Who are the people protesting against the Finnish coronavirus restrictions and spreading misinformation about the pandemic, masks, coronavirus tests and vaccinations? Spotlight joined coronavirus denier groups on alternative social

> media, and found links to both the US anti-coronavirus movement and the Russian Foreign Ministry.
>
> (YLE 31.5.2021)

In principle, the viewer is presented with many expressions in the ingress that fit into the framework of danger, such as protesters, spreading misinformation, coronavirus deniers as well as links to suspicious groups. The name "Pandemiankieltäjät" (pandemic deniers) is not specified during the programme. A citizen activist featured in the programme talks about their critical attitude towards changing the definition of a pandemic, but not towards the existence of the disease.

The journalists are seen in the position of investigative journalism, from which they examine the activist groups portrayed as suspicious. "They say they are ordinary concerned citizens who defend freedom, but the groups are spreading a lot of misinformation about vaccinations and the coronavirus." The programme explains that journalists have anonymously joined social media discussion groups to gather information, which as a method refers more to police infiltration of criminal activities.

Citizen activists and people participating in the protest who are the subject of the documentary are described with the following word choices: "pandemic deniers", "radicalised" and "noisy activists". The same target group includes "new age people", "alternative care advocates", "True Finns" as well as "anti-EU" and "far-right" descriptions. In addition, the activists portrayed are said to have links to international anti-corona movements, the Russian Foreign Ministry and anti-Semitism. Factors that unite the activists are said to be a belief in conspiracy theories and rebelling against elites and authorities.

Both the TV programme and the article material contain discourse about the suspiciousness of alternative views. Discourse of suspiciousness and danger is illustrated, for example, by the newspaper headlines: "The anti-vaccination movement is dangerous disinformation. The unvaccinated bear the responsibility for the suffering and death of loved ones. The coronavirus infection spread by the unvaccinated nurse led to a death in Eurajoki."Fear fuels the idea of polarisation, whereby the portrayed citizens are positioned in opposition to proper citizens.

By using certain keywords, concepts, symbols and stereotypical elements, things that are associated with them come to the public's mind, and this activates the interpretative framework. (Karvonen, 2000, p. 81–82; van Dijk, 2008, p. 220). The concepts and stereotypes in the examined material, such as "spreaders of misinformation and pandemic deniers", activate interpretive frameworks of conspiracies and danger. The documentary also features symbolic images taken from the websites of activists, such as memes associated with anti-Semitism, which activate the feeling of fear. The activist interviewed in the programme denies links to conspiracies, and talks about trolls who attacked the website and spread memes and hate speech.

Conspiracy theories are linked to people expressing alternative viewpoints in several of the material's media content without justifying the concept or this link. According to Kielitoimiston sanakirja, a conspiracy is "a secret agreement between two or more people, often for a criminal purpose" (Kielitoimistonsanakirja.fi). Conspiracy theories, on the other hand, generally refer to explanations that there is a conspiracy behind some event or process which is not shared in public (e.g. Hyvönen & Pyrhönen, 2023). By referring to conspiracy theories, scepticism or alternative views can be shown to be false and be positioned in the framework of danger.

Images of hospitals, crowds of people queuing for vaccines and protesters attached to news stories also show danger and fear. Images from intense care units of hospitals concretise the seriousness of the situation to the public.

Vaccine criticism in a frame of moralisation

> The anti-vaccine movement is about falling for conspiracy theories. It's reminiscent of getting mixed up with a cult. In both cases, basic scientific facts become meaningless to a person and are replaced with wrong information, disinformation. Suddenly, the world isn't round anymore but flat. This is what the anti-vaccine movement is about. […] The most important thing is to book and to get a coronavirus vaccination according to the instructions of your municipality.
> (AL 24.7.2021)

Pointing out right and wrong makes alternative views seem unnecessary and distorted. In this case, the subject is examined by criticising and correcting wrong views and by instructing the reader, as in the Aamulehti editorial above.

Pointing out right and wrong information and behaviour can be called moralising. According to the social psychologist Susanne Täuber, moralism is usually about recognising one point of view and directing others to think and act in the same way (Täuber, 2019, p. 172–173). In the media material, moralising takes the form of one-way mass communication, which can give the impression of mainstream media's attachment to authority communication. Moralising that persuades to hegemony appears in the article material, for example, as a request to take the vaccine according to authorities' recommendations or to limit the spread of the viewpoints of the infidel: "[…] we have to rein in the spread of misinformation even more" (AL 24.7.2021).

According to Täuber, moralisation has often been used as a strategic goal to get people to change their attitudes and behaviour (Täuber, 2019, p. 172). Moralisation creates behavioural norms and defines right and wrong actions. This kind of moralisation can lead to intolerance and a moral right to stigmatise others, which in turn causes polarisation and reduces social cohesion. (Täuber, 2019, p. 178.) Expressing alternative views through moralisation is problematic for democratic debate and produces a society that only accepts one truth.

From the perspective of polarisation, moralisation enforces juxtaposition (Brandsma, 2017; van Dijk, 2008). "Moralised persuasion creates new moral norms that lead to rigid categorisation". In this kind of a social environment, people are often easily divided into "morally good and bad". The situation causes stigmatisation and marginalisation of moral deviants. In this case, those who support moral norms are likely to discriminate against moral deviants and those who do not agree with the morals reject those who follow morals. (Täuber, 2019, p. 175–176.) Moralisation therefore has a negative effect on social cohesion and it does not promote the achievement of common goals.

There was some moralisation that increased juxtaposition in the media material. This was reflected, for example, in the fact that scepticism about the coronavirus vaccination was frowned upon, and people included in this group were publicly scolded. "There are vaccine-safety sceptics, alternative care advocates, government doubters, biblically fearful, pandemic risk deniers and conspiracy theorists. What would change their minds?" (HS 10.1.2021). References to the "pandemic of the unvaccinated" and blaming the unvaccinated for infecting or even the death of others were evidence of extreme moralising.

Rhetoric is based on a shared truth about a situation. According to Tankard (2001, p. 96–97), framing can be used to define a situation without the audience even realising that a moralising definition has been made. When vaccine-critical people are examined in the framework of moralising, their hesitation seems irrational, as in the following article by *Aamulehti* which was looking for reasons for staying unvaccinated:

> [...] It is clear that the risk of infecting others is much higher for the unvaccinated than the vaccinated. Why don't some people take the coronavirus vaccine, even if it prevents a serious disease and deaths?
>
> (AL 29.9.2021)

In the investigative online article of *Yle,* vaccine criticism was linked to economic benefits. The article points out that anti-vaccine advocates may be meriting from spreading their views, which can be seen as morally dubious. The article states: "The most active anti-vaccine advocates may benefit financially from spreading conspiracies" (YLE 17.2.2021). The beginning of the article refers to a study made in the United States, which found economic drivers. However, the mandate of the organisation is not disclosed in the article. The report was carried out by an organisation whose mandate is to conduct campaigns that weed out hate speech and misinformation, such as vaccine-critical communication (CCDH, 2022). The article describes a few government restrictions opposing Finns that accept donations or promote products. However, the article concludes that "the anti-vaccine movement doesn't seem to be becoming a significant business in Finland".

Alternative views in a frame of invalidation

Alternative views were not discussed reflexively in the material, but they were rather invalidated by intervening with and correcting the facts, as in the *Ilta-Sanomat* article "Nyt puhuvat he, jotka eivät vielä ole ottaneet koronarokotetta – asiantuntija vastaa 3 yleiseen huoleen" (Now those who have not yet taken the corona vaccine are speaking – an expert answers three common concerns) (IS 28.11.2021). Fact checking was most often carried out by expert statements from representatives of public organisations or by referring to a research consensus, such as: "according to research, the PCR test is reliable and according to research, masks, coronavirus tests and vaccinations are not harmful" (YLE 31.5.2021).

Moralisation and invalidation often concurrently framed the narrative. In the *Pandemiankieltäjät* programme, for example, moralisation is especially evident in pointing out right and wrong information. Simultaneously, incorrect information and alternative arguments are invalidated by expert statements. In the programme, an interviewee who is sceptical about coronavirus vaccines and who shares a view on the effect of lifestyle on getting sick with the coronavirus is corrected by the statement of a doctor who corrects alternative health claims.

Authoritarianism includes justification by speaker categories, whereby the speech of a respected expert, such as a known doctor, is seen as lacking alternatives and makes considering other positions unnecessary or even suspicious (Jokinen, 2002). The speaker categories in the media material do not include experts from fields such as virology or epidemiology who deviate from the communication of authorities.

The media material of the research included some media content where alternative views had been sought through public surveys and interviews. The surveys were often attached as part of reports carried out by media on the background of alternative views. For example, the *Aamulehti* survey involved readers by asking: "Tell us why you are not taking the coronavirus vaccine. Aamulehti is doing a story about it. Answer our survey and tell us your justifications." (AL 20.9.2021). An article following the survey reported that hundreds of people answered "the question why they are not taking the coronavirus vaccine, even though without it, they are at risk of serious illness and even death – in vain" (AL 29.9.2021). Possible justifications for refusing the vaccination were presented as insignificant from the point of view of a fatal illness. The article also explored the reasons for refusal by interviewing four people refusing the vaccination and four experts. Different views were expressed by involving readers and interviewing hesitant people, but simultaneously, alternative views were invalidated by facts according to official information, such as the fact that no widespread harms have come out from coronavirus vaccines: "The vaccines meet the criteria set for vaccine safety, and over six billion doses have been given. Monitoring data has not revealed any widespread harmful effects". These facts were placed in several paragraphs of the article after hesitant comments. A similar structure was repeated in several articles that were included and excluded from the analysed material.

Invalidation often refers to anti-science ideas and the falsehood of alternatives, as in the *Helsingin Sanomat* article, which discussed the spread of misinformation on social media, and portrayed people who trust in a healthy lifestyle as huggers who support alternative medicine:

> A peculiar campaign went viral on social media last November. Coronavirus can be avoided through a healthy lifestyle and by focusing on happiness, the campaign claimed. In the images, welfare sector influencers and promoters of alternative medicine were hugging each other in the woods. The claims were of course false.
>
> (HS 18.2.2021)

Conclusions

As the influence of media on society is obvious, and as media is seen as an important tool of influence in areas such as politics and civil engagement (Reunanen & Kunelius, 2021), whether citizens can make their views heard is essential for the functioning of civil society. A multi-voiced civil society raises important topics of debate for society and monitors the fairness of actions of decision makers (Harju, 2020; Järvinen, 2018).

Journalistic media often aims to produce multiple voices or to create a debate between different views. However, there are problems with multi-voiced debate – how to involve people expressing differing viewpoints and to secure respectful debate. From a journalistic storytelling point of view, it might be tempting to highlight extremes, through which the journalist portrays societal tensions (Hautakangas et al., 2017; van Dijk, 2008). This choice ignores the motives and underlying reasons behind the views, as well as the diversity inherent in multiple voices. The implementation of multiple voices by presenting extremes or opposite sides ignores the complexity of reality and can increase polarisation in communication.

This research was able to identify elements that enforced polarisation in communication in the media content of the early days of the pandemic, such as lack of dialogue, justification with speaker categories, categorising people as well as using extreme expressions. In addition, the material showed fact construction based on a selected moral code, which can increase polarisation between people. Negative stereotypes, black-and-white representations and rhetorical word choices are indicative of a polarised through structure.

Categorising and defining people from the outside causes tensions in communication between people. (Brandsma, 2017). In the media content, stigmatising and stereotypical expressions, such as: “conspiracy theorists”, “spreaders of disinformation”, “cultists” and “ignorant”, as well as generalising categorisations such as “anti-vaccine or anti-science advocates” could be categorised as fuel for polarisation.

The expressions, word choices, symbols and images found from the material activate interpretative frameworks of danger, moralisation and invalidation (see Table 8.1). These interpretive frameworks limit the public’s understanding of the diversity of situations. Division created by moralising stigmatises others who do not follow principles of practice. Thus, the communication climate also becomes one where keeping silent about difficult topics is a way of staying in the group of people “who are right”. Quantitative information of multiple voices could be found in the data, such as audience surveys and interviews with people with different views. However, the processing of this information led to the correction or invalidation of alternative or sceptical views.

The situation leads to reflection of the active involvement of journalism in the selection of certain viewpoints and the discrimination of others. Invalidation of alternatives can erode trust in media if they try to appear as a neutral actor at the same time. While the central task of mainstream media is to provide accurate

Table 8.1 Interpretative frameworks of diverging views on authority communication in the pandemic time media content, activated by word choices, symbols and images

	Danger	*Moralisation*	*Invalidation*
Word choices	Conspiracies Conspiracy theories Networking	Irresponsibility Hoax Misleading	Disinformation Misinformation Scam Anti-science
Symbols and metaphors	The day's corona figures Anti-Semitist memes	Police Infiltrator Journalists	Awakened activists Biblical reasons for refusal
Images	Graphics of infected and dead people Hospitals and intensive care units with patients People lining up for vaccinations	Crowds at protests Police at the protests People with corona	People celebrating and hugging
Identity speech about people expressing critical views	Conspiracy theorists Radicalised	Deniers Loud ones	The ignorant Tinfoil hats

Note: The interpretative frameworks often appeared side by side, whereby, for example, the public expression of different views was moralised and at the same time invalidated.

information, it should not be done by invalidating different views, let alone citizens expressing their concern. However, this appears to have happened, for example, with the coronavirus vaccine publicity, where sceptical citizens were often associated with anti-science or anti-vaccine ideas, which is not justified by research. Controversial topics, such as pandemic management, vaccine safety or, for example, alternative treatments, which can also include research data that is uncertain and hard to interpret, are difficult for journalists and require the ability to steer clear of juxtaposition that enforces polarisation, for instance by thinking about word choices.

Since the media has the power to define debate and its perspectives, the public adopts an attitude of discrimination from the speech it publishes (van Dijk, 2008, p. 55). Therefore, the stigmatisation, stereotyping and positioning itself as being morally correct by the mainstream media can contribute to the polarisation that is reinforced by identity speech. Should the journalistic media then take on the role of unravelling steepened polarisation or a bridge builder? Not necessarily, but if alternative views are stigmatised and invalidated, its societally defined role as a monitor of power use and facilitator of democratic debate will not be fulfilled. The role of media as a watchdog of power is at risk of being replaced by that of a gatekeeper of citizens.

Defending freedom of speech is an integral part of the societal tasks of journalism. The news media that were examined in this research are committed to comply with the journalistic guidelines and the principles of the Council for Mass Media in Finland. The Council for Mass Media in Finland monitors good journalistic practice and defends freedom of speech. According to the journalistic guidelines of the council, the human dignity of every person must be respected and minorities must not be portrayed in a degrading or derogatory light (JSN, 2022. In addition, the tasks of the public media are defined as promoting freedom of expression, pluralism and interactivity, which are mentioned in Finnish law and in the Treaty on European Union (Act on Yleisradio Oy 1993/1380 § 7; EUVL C 340).

Communication scholars have proposed socially responsible journalism and dialogical approaches to defuse tense communication. Socially responsible journalism can be linked to its societal task as a facilitator of democratic debate and as an actor following ethical guidelines. The aim of dialogical approaches is an equal and respectful debate (e.g. Ahva & Wiard, 2018; Heikka, 2021; Hautakangas et al., 2017). For example, in the story about the protest, rather than polarisation, it would be useful to listen to the diversity that can be found behind a seemingly unified group.

Disagreement is part of democratic debate culture, and persuading people to speak with one voice leads, at worst, to authoritarian thinking (Journalists Without Borders 2020; Baer & Koponen, 2020). To maintain social peace, one needs to be able to live and work with polarisation (Brandsma, 2017), which means tolerating alternative views and avoiding communication that stigmatises people. According to a report on Finnish media culture, the media should consider for whom and from whose perspective the content is produced, who remains invisible and in what kind of roles people from different backgrounds are portrayed (Kanerva et al., 2022).

Multiple perspectives are an important element in different kinds of crisis situations, where administrative communication meets the wretched challenges of a complex world. In exceptional circumstances, facts can be used to try to manage the uncertainty of citizens, but building trust and unity and tolerating uncertainty and different viewpoints is key. If the media is too obedient to the administrative agenda of media briefings, as often happens during a crisis, it might contribute to polarisation between groups of people and those with different views. Exceptional periods should not lead to a narrowing of media perspectives or to conforming to those in power.

A multi-perspective approach promotes active participation and enables the citizens' feeling of involvement, which can be seen as a prerequisite for social cohesion and a socially sustainable society. In order for a multi-voiced approach to relieve tensions, it should be respectful of all, non-judgemental and non-prejudiced. It would therefore be useful in media research to identify those journalistic interpretative frameworks through which the different views of citizens are interpreted as well as to consider the success of media as a facilitator of democratic debate. As a topic for further research, it would also be interesting to determine whether Finnish media coverage during the pandemic was more or less polarising than elsewhere.

The article is based on a research project called *Reilu media: Vaientamisesta kunnioittavaan moniäänisyyteen (2021–2024) (Fair Media, From Silencing to Respectful Pluralism),* which was funded by the Media Industry Research Foundation of Finland and implemented at Tampere University. The project examines the possibilities and obstacles of citizens' participation in media as well as develops ways to produce multi-voiced media content. The research consists of the analysis of media content published during the pandemic, cooperative workshops of journalists and citizen activists as well as interviews of media content producers.

Original article: Mäkinen, M. (2023). Kansalaisaktivistit vai pandemiankieltäjät? Vaihtoehtoiset äänet journalistisessa mediassa COVID-19-pandemian aikana. Media & viestintä, 46(3), p. 51–67. https://doi.org/10.23983/mv.137062.

References

Act on Yleisradio Oy (1993). Enacted in Helsinki 22.12.1993. Read 20.9.2021. www.finlex.fi/fi/laki/ajantasa/1993/19931380#L3

Ahva, L. & Wiard, V. (2018). Participation in local journalism: assessing two approaches through access, dialogue and deliberation. *Sur le journalisme, About journalism, Sobre jornalismo*, *7*(2), 64–79. https://revue.surlejournalisme.com/slj/article/view/359/349

AL (1.11.2021). Kallionpää, K. "If the vaccinated also get sick and spread the disease, does it make sense to get the vaccine at all? Expert: "The risk of a breakthrough infection is significantly reduced". Retrieved 4.9. 2023. www.aamulehti.fi/koronavirus/art-2000008374977.html

AL (24.7.2021). "The anti-vaccine movement is dangerous disinformation". Editorial. Retrieved 4.9. 2023. www.aamulehti.fi/paakirjoitukset/art-2000008145050.html

AL (29.9.2021). Ala-Heikkilä, M. " Why are some people not taking the coronavirus vaccine, even though it prevents serious illness and deaths? Aamulehti asked and hundreds answered". Retrieved 4.9. 2023. www.aamulehti.fi/koronavirus/art-2000008275646.html

AL (23.10.2021). Thomsson, I. "This is what it's like to treat corona patients in the intensive care unit at Tays – chief medical officer talks about serious cases and confusion: "The number of un-vaccinated people feels depressing". Retrieved 4.9. 2023. www.aamulehti.fi/koronavirus/art-2000008347339.html

Attias, M. & Kangasoja, J. (Ed.). (2021). *Me ja ne: Välineitä vastakkainasettelujen aikaan.* Into Kustannus Oy.

Baer, K. & Koponen, K. (2020). *Kiinan suuri harppaus: Päämääränä maailman valvotuin kansa.* Otava.

Brandsma, B. (2017). *Polarisation: Understanding the dynamics of us versus them.* BB in media.

CCDH (2022). *The Center for Countering Digital Hate.* www.counterhate.com/

Council for Mass Media in Finland JSN (2022). Journalistic guidelines. https://jsn.fi/journalistin-ohjeet

Entman, R. (1993). Framing: toward clarification of a fractured paradigm. *Journal of Communication, 43*(4), 51–58. https://doi.org/10.1111/j.1460-2466.1993.tb01304.x.

EUVL C 340. (2017). *Eur-Lex.* Euroopan unionin virallinen lehti. https://eur-lex.europa.eu/legal-content/FI/TXT/HTML/?uri=CELEX:12016M/PRO/29&from=EN

Farr, C. & Rodriguez, s. (2020). Facebook, Amazon, Google and more met with WHO to figure out how to stop coronavirus misinformation. *CNBC.* www.cnbc.com/2020/02/14/facebook-google-amazon-met-with-who-to-talk-coronavirus-misinformation.html

Fletcher, R., Cornia, A., & Nielsen, R. (2020). How polarized are online and offline news audiences? A comparative analysis of twelve countries. *The International Journal of Press/Politics, 25*(2), 169–195. https://doi.org/10.1177/1940161219892768.

Forsberg, T. (2020). Sodassa koronavirusta vastaan? Covid-19 ja metaforien pandemia. *Kosmopolis, 50*(2), 104–115. https://doi.org/10.1080/10926488.2018.1407992.

Goldenberg, M. (2021). *Vaccine hesitancy: public trust, expertise and the war on science.* University of Pittsburgh Press.

Hakala, S. & Ruggiero, A. (2022). Kriisinkestävään yhteiskuntaan. In M. Koskela & H. Kantanen (Eds.), *Poikkeuksellinen viestintä,* 28–48. ProCom.

Harju, A. (2020). Aktiivisen kansalaisuuden vaatimukset. *Kansalaisyhteiskunta.fi.* https://kansalaisyhteiskunta.fi/tietopankki/aktiivisen-kansalaisuuden-vaatimukset/

Harjuniemi, T. (2020). Koronakriisi ja liberaalin asiantuntijuuden kahdet kasvot. https://etiikka.fi/koronakriisi-ja-liberaalin-asiantuntijuuden-kahdet-kasvot/

Hautakangas, M., Ahva, L., & Haara, P. (2017). *Sovittelujournalismin käsikirja.* https://sovittelujournalismi.fi.

Heikka, T. (2021). Miten Minä ja Sinä voivat kohdata journalistisissa dialogeissa? *Media & viestintä, 44*(4), 20–45. https://doi.org/10.23983/mv.112866.

Hokkanen, A. (2009). *Susiriesa pian koko maassa: Suurpetojen representaatiot maaseudun elinkeinoihin keskittyvän erikoissanomalehden diskursseissa.* Master's thesis, University of Vaasa.

HS (10.1.2021). Nieminen, T. ”Yes I will say no”. Retrieved 4.9. 2023. www.hs.fi/sunnuntai/art-2000007727904.html

HS (18.2.2021). Salmi, S. Against scams. Retrieved 4.9.2023. www.hs.fi/elama/art-2000007809573.html

HS (18.5.2021). HS data desk. HS's updating graphics give the latest information about the coronavirus situation: 983 people in the hospital, 55 of them in intensive care. Retrieved 4.9.2023. www.hs.fi/suomi/art-2000006477656.html

HS (22.10.2021). Vasantola, S. ”Focusing only on disadvantages in the vaccination debate leads astray, for example, the risk of myocarditis is many times higher in this disease”. Retreived 5.9. 2023. www.hs.fi/kotimaa/art-2000008349675.html

HS (8.11.2021). Ala-Risku, P. HS's corona overview: ”The number of people in hospital on the rise again”. Retrieved 5.9. 2023. www.hs.fi/kotimaa/art-2000008389221.html

Hyvönen, A-I. & Pyrhönen, N. (2023) (Ed.) *Salaliittoteorioiden politiikat. Yhteiskuntatieteellisiä näkökulmia.* Vastapaino.

Ikäheimo, H-P. (2021). *Politiikkaa ilman vaihtoehtoja? Suomalainen sanomalehtijournalismi EU-keskustelun ja demokratian näyttämönä 1995–2018.* Doctoral thesis, Tampere University.

IS (21.10.2021). Lehto, M. ”President Niinistö in IS interview: ”The un-vaccinated have made themselves hostages”. Retrieved 5.9. 2023. www.is.fi/politiikka/art-2000008347999.html

IS (15.11.2021). Jääskeläinen, K. ”The un-vaccinated are overloading intensive care – Hus is already approaching a critical limit”. Retrieved 5.9. 2023. www.is.fi/kotimaa/art-2000008404615.html

IS (28.11.2021). Jääskeläinen, K. ” Those who have not yet gotten the coronavirus vaccination speak out – an expert answers three common concerns”. Retrieved 5.9. 2023. www.is.fi/kotimaa/art-2000008417912.html

Jallinoja, P. & Väliverronen, E. (2021). Suomalaisten luottamus instituutioihin ja asiantuntijoihin COVID19-pandemiassa. *Media & viestintä, 44*(1), 1–24. https://doi.org/10.23983/mv.107298

Järvinen, J. (2018). Kansalaisyhteiskunta: demokratian ytimessä vai periferiassa? In K. Miklossy & J. Nikula (Eds.), *Demokratian karikot: Itäinen Eurooppa suuntaa etsimässä* (pp. 163–174). Gaudeamus.

Jokinen, A. (2002). Vakuuttelevan ja suostuttelevan retoriikan analysoiminen. In A. Jokinen, K. Juhila & E. Suoninen (Eds.), *Diskurssianalyysi liikkeessä,* (pp. 126–159). Vastapaino.

Journalists Without Borders (2020). Lehdistönvapaus vaarassa koko maailmassa: Korona kiihdyttää sananvapautta uhkaavia kriisejä. www.toimittajatilmanrajoja.fi/post/lehdist%C3%B6nvapaus-vaarassa-koko-maailmassa-korona-kiihdytt%C3%A4%C3%A4-sananvapautta-uhkaavia-kriisej%C3%A4

Kampf, G. (2021). COVID-19: stigmatising the unvaccinated is not justified. *The Lancet, 10314*(398),1871. www.thelancet.com/journals/lancet/article/PIIS0140-6736(21)02243-1/fulltext

Kanerva, A, Lahtinen, E. & Oksala-Särelä, K. (2022). *Näkökulmia mediakasvatuksen ja medialukutaitojen yhdenvertaisuuteen*. National Audiovisual Institute (KAVI).

Kantanen, H. & Koskela, M. (2022). Poikkeuksellinen pandemiaviestintä. In M. Koskela & H. Kantanen (Eds.), *Poikkeuksellinen viestintä* (pp. 14–28). ProCom.

Karvonen, E. (2000). Tulkintakehys (frame) ja kehystäminen. *Media & Viestintä, 23 (2)*, 78–84.

Kemppe podcast (2021). Ihmisoikeudet, Pelolla Hallinta, Ylen Kuvaukset – Henna Maria #37. Listened 20.9.2021. www.youtube.com/watch?v=rXDKWrqDhks&t=12s

Kohvakka, R. & Saarenmaa, K. (2021). Median merkitys on kasvanut pandemian aikana – monet ikäihmiset ovat ottaneet melkoisen digiloikan. *Tieto & Trendit*. Tilastokeskus. www.stat.fi/tietotrendit/artikkelit/2021/median-merkitys-on-kasvanut-pandemian-aikana-monet-ikaihmiset-ovat-ottaneet-melkoisen-digiloikan

Kunelius, R., Noppari, E. & Reunanen, E. (2010). *Media vallan verkoissa*. Tampere University

Lehtola, V-P. & Autti, O. (2019). Jurnutuksen käsikirja. In O. Autti, & V.-P. Lehtola, (Eds.), *Hiljainen vastarinta*. Tampere University Press.

Mallon, M. & Saalo, V. (2021). Kriittinen journalismi ja poliittinen pelko. In *Koronan kääntöpuoli*. Helsinki/Porvoo: Särötär.

Matikainen, J., Ojala, M., Horowitz, M. & Jääsaari, J. (2020). *Media ja yleisön luottamuksen ulottuvuudet: Instituutiot, journalismi ja mediasuhde*. University of Helsinki.

Mattila, N. (2022). *Heränneet. Maailma salaliittoteoreetikoiden silmin*. Otava.

Metelinen, S. (2021). *Kansakunnan kivijalka*. EVA analysis No. 95/2021.

News Media Finland (2022). *JSN ja Journalistin ohjeet*. www.uutismediat.fi/sanomalehtitieto/journalismi/

Niemi, M. K. (2022). Hylkääkö media polarisoituvat yleisöt? Comment in Reunanen, E., Alanne, N., Helske H., Lappalainen, E., Niemi M.K., Pettersson, M. & Seur, V. (eds.) *Uutismedia verkossa. Reuters-instituutin Digital News Report. Finland's National Communication*. Tampere University.

Niemiec, E. (2020). COVID-19 and misinformation: Is censorship of social media a remedy to the spread of medical misinformation? *EMBO Reports*. Nov 5, *21*(11). www.embopress.org/doi/full/10.15252/embr.202051420

Noelle-Neumann, E. & Thomas P. T. (2004). The spiral of silence and the social nature of man. In L.l. Kaid. (Ed.), *Handbook of political communication* (pp. 339–356). Lawrence Erlbaum.

Nurmi, J. (2021). Building 'natural' immunities: cultivation of human-microbe relations in vaccine-refusing families. In C. Brives, M. Rest & S. Sariola (Eds.), *With Microbes* (pp. 104–121). Mattering Press.

Nurmi, J. & Vuolanto, P. (2020). Tutkimukseen perustuvia näkökulmia rokotuskeskusteluun. *Sosiaalilääketieteellinen aikakauslehti 57* (3), 213–214.

OECD (2021). *Civic space scan of Finland.* OECD public governance reviews, OECD Publishing. https://doi.org/10.1787/f9e971bd-en

Ojala, M. (2021). Is the age of impartial journalism over? The neutrality principle and audience (dis)trust in mainstream news. *Journalism Studies 22*(15), 2042–2060. https://doi.org/10.1080/1461670X.2021.1942150

Pitkänen, T. (2021). *Keitä ovat nämä kaiken maailman pahansuovat tietolaarinvartijat*? https://puheenvuoro.uusisuomi.fi/tarjapitknen/keita-ovat-nama-kaikenmaailman-pahansuovat-tietolaarinvartijat/

Reiter-Haas, M.; Klösch, B; Hadler, M. & Lex, E. (2022). Polarization of opinions on COVID-19 measures: integrating twitter and survey data. *Social Science Computer Review*. https://journals.sagepub.com/doi/epub/10.1177/08944393221087662

Reunanen, E. (2022). Poliittinen polarisaatio ja journalismi: neljä medialuottamusprofiilia. *Media & Viestintä 45*(4), 22–45. https://doi.org/10.23983/mv.125625.

Reunanen, E., & Herkman, J. (2020). Viestinnälliset oikeudet tutkimuksen polttopisteessä. *Politiikasta.fi.*

Reunanen, E. & Kunelius, R. (2021). *Medioitunut valta ja politiikan paluu.* Survey on Finnish decision-makers' relationship with the media in 2009 and 2019. Tampere University Press.

Schmidt, H. (2022). Pandemics and politics: analyzing the politicization and polarization of pandemic-related reporting. *Newspaper Research Journal*, *44*(1). https://doi.org/10.1177/07395329221095850

Siippainen, A. (2021). Nina Honkanen on mukana lääkäriryhmässä, joka vaatii lasten koronarokotusten lopettamista: Näin hän perustelee kantaansa – "Missä on kriittinen tieteellinen keskustelu?" www.lansivayla.fi/paikalliset/4286380

Simonen, J., Westinen, J., Pitkänen, V. & Heikkilä, A. (2021). *Luottamusta ilmassa, mutta kuinka paljon?* A study on the trust of different generations in the institutions of society. E2 Research.

Smith, S. (1997). "Developing new reflexes in framing stories". In *Pew Center/RTNDF workshop civic journalism: doing it daily*, Chicago.

Tankard, J.W. (2001). The empirical approach to the study of media framing. In S. Reese, O. Gandy & A. Grant (Eds.), *Framing public life. perspectives on media and our understanding of the social world.* Routledge.

Täuber, S. (2019). Moralisation as legitimation for ostracism. In S.C. Rudert, R. Greifeneder, & K.D. Williams (Eds.), *Current directions in ostracism, social exclusion, and rejection research* (pp. 171–189). Routledge. https://doi.org/10.4324/9781351255912.

Tilastokeskus (2022). *Kansalaispulssi* 27.1.2022.

Tolonen, E. (2021). Journalismi, hallituskriisit ja polarisaation ilkeä kierre. *Politiikasta.* https://politiikasta.fi/journalismi-hallituskriisit-ja-polarisaation-ilkea-kierre/.

Väliverronen, E. & Ekholm, K. (2020). Tieteen vapauden ja tutkijan sananvapauden yhteiskunnallinen merkitys. In E. Väliverronen & K. Ekholm (Eds.), *Tieteen vapaus & tutkijan sananvapaus*. Tampere: Vastapaino.

Väliverronen, E. & Jallinoja, P. (2021). Suomalaisten näkemykset asiantuntijoista ja rokotuksista. Havaintoja muutoksista ensimmäiseltä koronavuodelta. *Yhteiskuntapolitiikka, 86*(3), 323–333.

Van Dijk, T. A. (1993). Principles of critical discourse analysis. *Discourse & Society, 4*(2), 249–283.

Van Dijk, T. A. (2008). *Discourse and power.* Hampshire: Palgrave Macmillan.

YLE (17.2.2021). Rummukainen, A. "Yle found out: This is how the anti-vaccine movement raises money in Finland – including a convicted financial offender". Retrieved 5.9. 2023. https://yle.fi/uutiset/3-11772207.

YLE (31.5.2021). *Pandemiankieltäjät.* Spotlight. Yle Teema Fem. Retrieved 5.9.2023. https://areena.yle.fi/1-50677547.

YLE (21.10.2021). Laakso, A. Yle's information: The coronavirus infection spread by the unvaccinated nurse lead to a death in Eurajoki – chief medical officer wants vaccines to be required for nurses. Retrieved 5.9.2023. https://yle.fi/uutiset/3-12153343

YLE (2022). Toimintaperiaatteet.. https://yle.fi/aihe/yleisradio/toimintaperiaatteet.

Part 3

Constructing public knowledge and trust

9 Mediated science and issues of public knowledge and trust

Anna Maria Jönsson

Introduction

We live in a world of wicked problems, and science has become increasingly vital for public life, politics, and individual decision-making. However, science, along with other knowledge institutions, like journalism, is facing growing scrutiny and questioning. Contemporary public discourse often grapples with issues involving fake news and post-truth narratives. Consequently, the topics of public knowledge and trust in science are of the utmost importance. Regarding public knowledge, the media serves as a fundamental platform for public discourse, shaping what the public considers to be important and whom and what to trust. Media outlets also provide information about scientific topics and influence public perceptions of which issues and disciplines are of particular significance (Bauer et al., 2006; Nelkin, 1995). Generally, media content is viewed as both reflective of, and influential in, shaping public opinion (Pellechia, 1997). Therefore, how science is portrayed in the media is a crucial consideration when studying science communication (Schäfer, 2012).

The media and science have an important reciprocal relationship, despite operating under different principles (Olausson & Berglez, 2014; Schäfer, 2017). In recent decades, there has been increased interest among researchers in communicating through the media (including traditional news media, as well as social media platforms), which has been paralleled by growing media interest in reporting on scientific research (Rödder & Schäfer, 2010). This increasing interrelation between science and the media has been described as a mediatisation of science (cf. Rödder & Schäfer, 2010). According to Sharon Dunwoody (2021), science journalism has never been more important than it is now. The media serves as a bridge between scientific research and citizens, providing researchers with a window through which they can engage with the public. For the media, science is not only a subject of news but also a valuable resource for credibility and factual information as discussed by Abalo in this volume (Abalo, 2025). In many cases, the media has been seen to monopolise communication and act as the primary source of information for citizens (Schäfer, 2012).

All science communication relies heavily on trust. Aside from knowledge, values, ideologies, and identities, the media is one of the most important factors influencing

DOI: 10.4324/9781003479550-12

public trust in science (Roberts et al., 2013; Schäfer, 2016). As Goldenberg (2023, p1) has stated, "…public trust in science and/or scientific institutions is crucially important for science and society." Trust is the glue of society and influences more or less every kind of human interaction. Furthermore, trust is a precondition for the successful dissemination and use of knowledge in public deliberation and policy making. Traditionally, the media in this context has referred to news media, with most studies having been conducted on newspapers. Even though a significant amount of science communication today flows through a hybrid media system (Chadwick, 2017), this chapter will continue with this focus on the news media.

A number of studies from the last couple of decades have confirmed the continuous increase in the media coverage of science in different countries, and this increase has been particularly evident in print media (cf. Bucchi & Mazzolini, 2003; Pellechia, 1997; Schäfer, 2011). Previous research analysing media representations of science has mainly focused on science as a technology, the natural sciences, and certain research areas and issues of special interest. There are, however, some exceptions (e.g. Summ & Volpers, 2016). Against this background, this chapter aims to provide a more comprehensive analysis of how research and science are presented in news media discourse, with the further objective of allowing this analysis to feed into discussions on the relation between media discourse and public trust in science. This means that both the concepts of science and research will be used, since the concept of science in some other contexts refers primarily to research being conducted in the STEM[1] area.

In the situation with resistance, post-truth and filter bubbles, not only science but also the role of the news media and journalism has been questioned as an expression of and a consequence of increased ideological polarisation (Andersson et al., 2017; Dunwoody, 2021; Nichols, 2017). The polarisation in terms of trust in knowledge institutions like science and media is not least prominent in relation to current societal crises as climate change and the Covid pandemic. There was consequently also an increased interest in the research field in relation to the Covid pandemic when issues about trust and science came to the fore both in research and public debate (Agly, 2020; Bromme et al., 2022; Petersen et al., 2021; Sulik et al., 2021).

Following Schäfers' (2016) and Weingarts' (2022) arguments that trust in relation to science is an important category for analysis of contemporary society, that trust in science to a large extent is the result of mediated communication and that different form of mediatisation could affect public trust in science, the aim of this chapter is to analyse public communication and images of science. The chapter focus on Sweden as a case and will empirically answer the following questions:

1 How is science and research represented in Swedish news media in terms of frames and actors and are there any differences between different disciplines?
2 How do researchers communicate with media and what is the role of journalists in framing news about science and research?
3 What are the trends in levels of trust in science and research over time, and what factors might explain variations in trust among different demographic groups within the population?

The research questions are explored using a mixed-methods approach that combines qualitative interviews with quantitative content analyses and survey data. All empirical findings are drawn from earlier studies conducted between 2018 and 2021, which have already been published in Swedish. This chapter aims to synthesise these findings addressing the over-arching question of the role of media in the relationship between science and the public.

Mediated information about science is crucial for public life and citizens need to get access to research and science but also to understand it, and trust in the sender of the information is important for how it is used. That today science isn't only mediated but there is an increasing mediatisation process going on could possibly affect the relationship between public and science in different ways. On the one hand scientists are adjusting to the news media format in their communication which can improve the accessibility of the information for the public, but on the other hand mediatisation of science is also closely connected to the professionalisation of science communication since a higher communicative competence is needed. A situation where scientists themselves are not the sender of the information but instead communication and PR experts may affect trust negatively. In relation to this mediatisation of science thus can be considered a double-sided coin as it could both be a facilitator of as well as a hindrance to public trust in and knowledge of science (cf. Weingart, 2022).

After this introduction the chapter presents the two main theoretical perspectives which are *Science and media* and *Science and trust*. In the next section the method and material for the analysis is presented and followed by the analysis. The analysis section is structured in line with the three research questions. A concluding discussion ends the chapter.

Science and media

There has been a shift in perspective in science communication research from seeing media as a mediator of information from researchers to citizens, to discussing different forms and levels of mediatisation of research and science[2] (cf. Schäfer, 2009; Comfort, Gruszczynski, & Browning, 2022; Weingart, 2022). Within the research area of science communication, the concept of mediatisation has been defined, among other things, as extensive or increasing content; an increasing degree of pluralisation in terms of subjects and actors; as well as an increasing degree of conflict orientation in the reporting of science (Schäfer, 2009). However, the mediatisation has seemed less clear in science compared to other areas of society (e.g. politics, sport and religion) (Rödder & Schäfer, 2010).

Rödder and Schäfer (2010) have studied what they call the macro level of mediatisation – the relationship between different social systems – with a focus on the relationship between science and the media. Among other things, they state that the scientific sphere differs from several other spheres of society (e.g. politics) in that it is not the general public that is the primary audience and that it has its own "public" in the form of the intra-scientific peer review system. However, the general public is becoming increasingly central as recipients of and participants in science communication and it has for different reasons become increasingly important for researchers to communicate with the surrounding society. This kind of communication is for example frequently required by funders (Dudo, 2015; Davis & Horst, 2016; Weingart, Engels, & Pansegrau, 2000; Weingart & Guenther, 2016).

Science is communicated to the general public in different forms like museums and public lectures, but there is no doubt that media is one the most important platforms. This means that science communication to a large extent is adjusted to media logics and that journalists are important for how science is framed and represented. Recent shifts in the media ecology affects patterns of media use as well as the terms for science communication and the relationship between scientists and journalists (Dudo, 2015; Dunwoody, 2021).

Traditionally the relationship between researchers and the news media has followed the same pattern in terms of power relations as previously the relationship between politicians and the media. Before the field of science journalism matured and became its own genre with specialists among the journalists, the media more or less reported what the researchers communicated. The science expert was clearly in charge. With the increasing mediatisation of science researchers have expressed concerns that they are losing their status as experts and that their authority is being questioned (cf. Rödder & Schäfer, 2010; Weingart & Guenther, 2016). With increasing digitalisation and the formation of a hybrid media system power over science communication shifts again, allowing for researchers as well as general public to themselves produce media content, like for example lectures and Ted-talks on Youtube (Dunwoody, 2021). Weitkamp et al. (2021) however conclude that traditional news media and journalists still play a very important role for public science communication.

To communicate relevant research and science to the public in an understandable way is not an easy task and the quality of science journalism has been questioned and journalists who work to report on research and science have been criticised. The critique comes from two sides – on the one hand, they have been considered to be too poorly trained in the conditions of research; on the other hand, those who actually are educated scientifically are seen as thereby adopting the researchers' perspective and losing their ability to be critical and scrutinising. According to the critique, the media also often omit information about science as a process and, for example, methodological issues, something that has been seen to affect public trust in science negatively (Mamboleo, Chebutuk, & Matu, 2023; Pellechia, 1997). There is also what can be called a dualism regarding news about research. On the one hand journalism is seen as a popularisation of research (results) and on the other hand research is used as a source/perspective for other news (e.g. about politics, crime, sports etc).

Previous research show that in line with the mediatisation discourse, the extent of news about research and science generally have increased over time in terms of the number of articles and the articles about research also have tended to be longer (Bucchi & Mazzolini, 2003; Pellechia, 1997). Also, previous research focuses on different research areas, but the overall conclusion about research in the news media is that it is news in the medical field that dominates, and this applies both in Sweden and internationally (Pellechia, 1997; Vetenskap och Allmänhet, 2014). Different research questions and results receive attention to varying degrees and it is usually only the final phase of the research that is of interest to the media. Research that fits media logic and news evaluation naturally gets the most space (Rödder & Schäfer, 2010).

In journalism studies conflict is usually highlighted as a one of the most central news values (e.g., Galtung & Ruge, 1973; Harcup & O'neill, 2017). Previous research in the field of science journalism suggests that news about research and science differ somewhat in this respect and that conflict regarding research results etc. only leads to uncertainty among journalists as well as the public. Instead, it is the relevance of the research to topics that receive great attention at a certain point in time that is decisive for news value (Bucchi & Mazzolini, 2008). Bucchi and Mazzolini (2008) also studied whether the portrayal of science in journalism tend to be framed as positive, negative, or neutral. Based on an analysis of the Italian daily press they conclude that approximately half of the articles presented research in a positive light, while about a quarter took a negative perspective, emphasising risks, and another quarter remained neutral. This study also revealed a dominance of international news coverage, particularly from the USA. However, national news also gained significant attention in the media's coverage of science.

Another central question concerns who is allowed to speak in journalism about research issues. Previous research shows that when science is discussed in the media, it is not infrequently actors other than researchers who are interviewed or used as sources in other ways. It is also shown that it is often a smaller number of researchers who are given space in the media. These are researchers with high positions and who do well in the media and master media logics, which means that this follows the same pattern as for journalism in general. Experts and researchers however get increased visibility and space both as sources, interviewees and as authors of opinion pieces (Rödder & Schäfer, 2010; Vetenskap & Allmänhet, 2019c). Science journalists are in general dependent on scientists as expert sources in what has been described as "passive newsmaking" (Comfort, Gruszczynski, & Browning, 2022). A study of Italian news media between 1946–1997 showed that about a quarter of all news is the result of active action on the part of researchers – for example in the form of press releases (Bucchi & Mazzolini, 2003) and in other cases the researchers are allowed to speak and answer the journalists' questions. In this way, it is possible to distinguish between researchers as reactive and proactive sources of news about research. According to Schäfer (2011) the most common source for science journalists traditionally has been a few selected influential scientific journals. However, digitalisation processes have provided new channels through which science content can be communicated to the public, by researchers as well as by professional science communicators and journalists (Ginosar, Zimmerman, & Tal, 2024).

A study of how research and science have been represented in the Swedish context between 2002 and 2013 among other things shows that there is a variation in the number of articles across different years, with medicine and social sciences being prominently featured, whereas humanities and educational sciences are notably underrepresented. Additionally, the most common form of coverage is general news articles, and only a lesser part consists of opinion pieces and editorials. It is noteworthy that approximately half of the articles analysed, focus on research findings and reports (Vetenskap & Allmänhet, 2014).

In summary, the interest in how the news media reports on research and science has been quite extensive. However, the focus has been on how science (i.e., natural science, technology, and medicine) is reported and the analyses often focus on specific research areas (such as the mapping of the human DNA), while studies of how the social sciences and humanities are represented are considerably less common. Most studies also focus on one country (usually in the West) and on the daily press. The proportion of longitudinal studies has increased over time from the 1960s onwards (cf. Schäfer, 2009, 2017).

What distinguishes this study in relation to previous research is that it analyses research in general and includes all disciplines, i.e., also social sciences and humanities. Admittedly, this was also done by Bauer et al. (1995), in what must be considered one of the most ambitious studies on research reporting that has been done but is otherwise relatively unusual (see e.g. Bucchi & Mazzolini, 2003). However, this study follows the pattern in previous research in that it mainly focuses on the daily press. The study is also longitudinal.

Besides from having access to relevant and correct information about research and wicked problems in society, trust in researchers and research institutions is an important factor influencing public willingness and abilities to participate in public life.

Science and trust

Trust has been defined and studied in a number of different ways and in different disciplines. Trust in science and research is mainly about what can be called epistemic trust: i e trust in the competence and knowledge produced by scientists (Sulik et al., 2021). This kind of trust is of relevance in situations where individual decision-making and behaviour is influenced by experts. Epistemic trust requires ongoing efforts to build and maintain trust through transparency, accountability, and education (Kitcher, 2011; Hendriks, Kienhue, & Bromme, 2016).

Trust is also about relations and the perspective considered most relevant for this study deals with relationships between subjects (individuals) and various political objects (Gamson, 1968). Previous studies have shown that interpersonal trust is important for the degree of trust in society's institutions. Often, however, the concept of trust in social science research is used to describe interpersonal relationships (Rothstein, 2005).

Previously, the focus in research on the relationship between research and citizens was on solving the problem that citizens had too little knowledge and too negative a view of science (science literacy model, public understanding of science), but then has moved on to instead highlight the importance of trust and including the citizens in the research and have a dialogue about research issues and results (Schäfer, 2009). To empirically analyse the relation between science and citizens (from public understanding of science to public attitudes towards science to public trust in and relationship to science), surveys have been the main tool. UK and USA have been the forerunners but now there are global surveys as well, like Wellcome Global Monitor from 2019 (Goldenberg, 2023) and studies in several other countries like Sweden (Norén Bretzer, 2017).

Citizens' trust in various social institutions and actors thus have been analysed in international studies and in Sweden by the SOM-institute for many years and from many perspectives. Trust as a phenomenon, can be described in three different levels: idea, institution and actor (Norén Bretzer, 2017; Norris, 2011; Peters, 2002) and there is clearly a difference between trust in science as institutions and scientists as individuals. Previous research has shown that trust in institutions generally tends to be higher than in individual actors, but when it comes to the relationship between universities and researchers, the situation has been the opposite, i e that trust in researchers has been higher than trust in universities (Norén Bretzer, 2017). Moreover, in studies of trust in science, "science" is generally used as a generic concept and does not take into account possible differences between different disciplines. There are however exceptions and Funk (2017) e.g., concluded that in the USA public trust in scientists generally is higher compared to other groups in society, but that there is a difference between different research fields – medical scientists are e.g. more trusted than climate scientists. Funks' study however only focused on STEM-research.

The degree of trust in research in general and the view of different research disciplines and research priorities differ between different groups in the population (Achterberg et al., 2017; Bergström & Oscarsson, 2015; Holmberg & Weibull, 2013; Norén Bretzer, 2017; Rothstein, 2005). These patterns can be found in Swedish as well as international studies. To a large extent this is the same pattern found in studies of citizens' trust in institutions and actors in society in general. In research, various factors have been used to explain these variations in levels of trust between different groups of the population. The factors that have proven to be most significant are education, age, proximity to research and higher education, as well as party sympathies (Achterberg et al., 2015; Bergström & Oscarsson, 2015; Holmberg & Weibull, 2013; Norén Bretzer, 2017; Rothstein, 2005). Proximity refers on the one hand to a geographical dimension where it is about whether one lives near a university and on the other hand a social and cultural proximity which is instead about the level of education and whether one has experience of studying at a university. The degree of trust is thus influenced by group affiliation and background, one's own identity and values, but also by the information one receives about the institutions and actors and the image one has of what they do and how well they carry out their mission. Here, reporting in the news media is of central importance and thus trust in media is also a factor to be considered addressing the relationship between science and the public. Studies of media trust in Sweden shows that it in many ways follows the same patterns as trust in science and that there are increasing trust gaps linked to political party affiliation. Among those with the least trust are right-wing sympathisers, who tend to hold a generally critical view of Swedish news media as well as towards research (Andersson, 2019; Norén Bretzer, 2017).

Citizens' idea of research is affected by the media's reporting and how the media choose to report can lead to both increased and decreased trust and different views on what is important and should be prioritised (Vetenskap & Allmänhet, 2011). Studies have shown that citizens' trust is in any case to a certain extent connected to the extent to which the media report on cheating and misconduct in research (Holmberg & Weibull, 2013) and scandals will affect trust in science and scientists negatively (Goldenberg, 2023). The media reporting surrounding the "Macchiarini scandal" when the work of the renowned surgeon Dr. Paolo Macchiarini, came under scrutiny and revealed allegations of scientific falsification and unethical practices, for example, came to affect Swedish citizens' trust in research in general and medicine in particular, but it was also clear that this decline in trust was only temporary (Vetenskap & Allmänhet, 2018). In the long term, negative reporting on research in the media does not seem to contribute to reduced trust levels among the audience but trust is at the same time dependent on the research being seen in the media – familiarity is one of the prerequisites for trust. The media image also affects which issues citizens think are important (cf. McCombs, Maxwell, & Shaw, 1972).

During and after the Covid pandemic there have been an increase in studies about trust in relation to science (Agly, 2020; Bromme et al., 2022; Petersen et al., 2021; Sulik et al., 2021). As a central part of providing citizens with expert knowledge the role of and trust in (news) media has been acknowledge in relation to the pandemic as well as to other societal crises (cf. Stiernstedt, 2021). Several studies show that people with a high level of trust in research and science were reportedly more likely to comply with official guidelines (Bicchieri et al., 2021; Petersen et al., 2021; Stosic et al., 2021; Turska-Kawa et al., 2022). Sulik et al., (2021) on the other hand come to the conclusion that this is a simplified picture and that trust in guidelines does not necessarily equals trust in science.

Science and scientists were given an uncharacteristic visibility in media during the pandemic and it is also shown by numerous studies that there was an increase in public trust in the beginning of the pandemic. In some countries the level of trust then decreased again but was generally found to be higher than before the pandemic (cf. Bromme et al., 2022; Goldenberg, 2023; Knudsen, Nordø, & Iversen, 2023). Bromme et al. (2022) for example studied if and how trust in science changed in Germany from before to during the pandemic. Their analysis shows that there was an increase in trust in science in the beginning of the pandemic and although it slightly declined thereafter, the level of trust was still higher than before the pandemic. Moreover, it is concluded that levels of and patterns of change in trust is related to education and political preferences, and that trust was closely related to expectations about how politics should handle the pandemic.

Method and material

This study of mediated science and trust empirically focus on Sweden as a case study. Swedes generally have a high level of trust and traditionally a high news consumption, but according to previous research when it comes to media representations of science and research and trust in science, more or less the same patterns can be found in Sweden as in other parts of the Western world.

The result section in this chapter builds on different studies conducted during the period 2018–2021 in the context of the over-arching research program "Science in Society" and "Communication about Corona" led by the Swedish organisation Public & Science Sweden.[3] The results have been published previously in separate reports and mainly in Swedish. This chapter aims to synthesise the results and conclusions from these different studies. The research questions will be answered by a mixed-methods approach which is needed for this complex area of study.

More precisely the results are based on four different material:

1 Quantitative content analysis on (a) general news about research in Swedish newspapers including the dominating national broadsheets (4) and tabloids (2) as well as the dominating local broadsheets (10) between 1995–2015, (b) opinion pieces in the Swedish broadsheet Dagens Nyheter during 1992–2015
2 A study of trust and media (survey studies as well as content analyses) in relation to the Covid pandemic (2020–2021)
3 Interview studies with Swedish journalists (2019) and the public (2015)
4 Survey studies of the Swedish population, mainly data from the Swedish SOM-institute[4]

Results/analysis

Science and research in in Swedish media

The study of the representation of research and science in Swedish newspapers during the period 1995–2015 (measuring every fifth year) included 1,764 articles which means between just under 300 and 500 articles per year distributed in the following way:

- 1995: 264
- 2000: 488
- 2005: 379
- 2010: 296
- 2015: 336

During this period the reporting of research and science in the Swedish daily press was stable with no real patterns of increases or decreases to be found. The length of the articles was around 400 words on average. It is quite possible that the study includes too few impact years or days to identify fluctuations here, but neither the number of articles nor their size is significantly different between the two extreme measurement points of 1995 and 2015, respectively (Vetenskap & Allmänhet, 2019c).

Previous research has shown that medicine dominates news reporting, followed by social sciences. This is partially confirmed by this study, where roughly 60% deals with one of the research areas of social science (33%) or medicine (28%). The humanities follow next (17%) and least amount of space is given to news in the field of educational science (2%) (see Figure 9.1).

Compared to previous studies (Vetenskap & Allmänhet, 2014), the humanities in particular seem to have increased their share over time, although a large decline for the humanities in 2015 somewhat breaks the pattern. Also, the share of news about social science is increasing over the years. However, as the study includes relatively few impact years and a limited selection of articles each year, no reliable conclusions about changes over time can be drawn (Vetenskap & Allmänhet, 2019c).

Of all the news articles about research and science only a small part (7%) is published in the special science sections that some (mainly national broadsheets) newspapers have. Science and research news is instead mainly found among local news (27%) and on the pages with domestic (mainly political) news (14%), as well as in the foreign politics pages and in the culture section. The articles that are selected for the special science sections are mainly about medicine and natural sciences. Together these categories account for about three-quarters of these news. News about the humanities dominates among the news about research and science in the culture pages, where the articles also are somewhat longer. News about

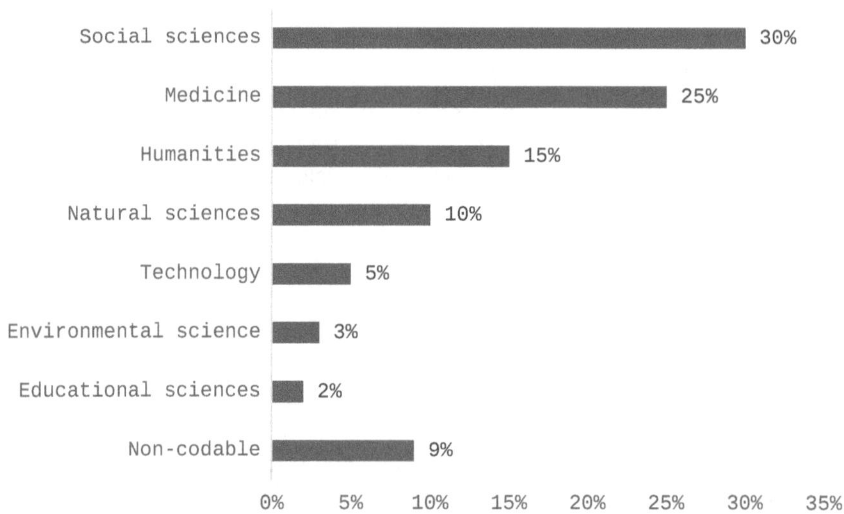

Figure 9.1 News article distribution per research discipline in Swedish press 1995–2015.

medicine and social sciences is usually placed among local news and often linked to a university in the immediate area. As for changes over time in the genres and forms of news about science and research it is difficult to discern any real trends, but the proportions of debate articles and articles in the cultural sections seem to be slightly increasing (Vetenskap & Allmänhet, 2019c).

News about research and science were framed in several different themes with health being the largest category with just over 20% of the articles, followed by culture (14%), research (12%) and politics (9%). Another aspect of how media frame news is the question of whether the news media use a positive, neutral or a negative frame of research and science. Here the results show that 49% of the articles are neutral and present the research without either a positive or negative angle, while around 40% of the articles present the research from a positive angle. Only about 10% of the articles use a negative frame. That journalists use a critical or negative frame is most common in articles in the science and editorial sections. News about medicine, science and technology are more often than news in the social sciences and humanities reported with a positive frame. Science journalism has traditionally been primarily of an informative and enlightening nature, focusing on research results, while investigative journalism has been given less space (see e.g. Bauer & Bucchi, 2007), a conclusion which is also confirmed by this study (Vetenskap & Allmänhet, 2019c).

> Example of positive frame:
>
> The dream – to be able to fall asleep. Every other woman over 50 have sleep problems. /.../ Women over 50 sleep the worst and eat mostly sleeping pills. Therefore, the Knowledge Center for sleep-disorders in Uppsala started a large investigation where 1500 women are interviewed about how they sleep. /../ To the comfort of the sleep-disordered, she believes (ref to Harriet Bengtsson, doctors) on a breakthrough within the next ten years period. (Aftonbladet/tabloid, 2000-03-20). Excerpt from article. Translated with google translate.
>
> Example of negative frame:
>
> Psychiatric drugs cause much of the senseless violence. During 1994, the question of where the senseless violence comes from have really been actualized. Psychologists talk about the father role. Psychiatrists talk vaguely about some place in the brain. It's really just psychiatric smoke curtains for the real reasons. /.../ (Helsingborgs Daglad/local, 1995-01-20). Excerpt from article. Translated with google translate.

Another aspect of how research and science is framed in news media concerns whose voices are being heard. As stated above, previous research shows that it is often other actors than researchers who are interviewed and quoted and that it is a smaller number of researchers – then mainly those with high positions – that is being allowed to comment these news (Ginosar, Zimmerman & Tal 2024;

Rödder & Schäfter, 2010). There however seem to be a trend that experts and researchers get increased space both as sources, interviewees and article authors and in a study of Bucchi and Mazzolini about a quarter of all news could be said to be the result of active action on the part of researchers and/or their employers (eg. communications departments) – for example in the form of press releases (Bucchi & Mazzolini, 2003). Besides from this, researchers were allowed to speak and answer the journalists' questions.

In this study about 35% of the actors who speak out or are mentioned in the articles are researchers or professors, generally affiliated to a university. The articles are also dominated by men, and only just under 30% are women. The male dominance is most evident among news about STEM-related research (medicine, natural science and technology). Besides from being a result of media logic and news values, whose voice that is being heard in news media is also affected by the communication steps taken by the different actors (Vetenskap & Allmänhet, 2019c).

Scientists and researchers as communicators and sources

That Swedish researchers have an obligation to engage with society at large is clearly stated within the framework of higher education regulations. Disseminating their research through various media platforms constitutes one important way for researchers to do this. Among these platforms, the news media stands out as a pivotal arena for the convergence of research and public discourse, but typically this arena is controlled by journalists. However, in op-ed sections publishing opinion pieces, researchers themselves – to varying degrees – can communicate directly with the public. Consequently, to publish opinion pieces serves as a means for researchers to connect with media audiences while retaining considerable control over content and language, albeit within the editorial oversight exercised by editors who select topics, authors, and partially guide article formatting, contributing to the mediatisation of science communication. To communicate and be seen in public platforms can also be part of a strategy for the researcher for increasing the chances of receiving funding for research.

One very important and culturally significant platform for public discourse in traditional news media in Sweden, is the op-ed section in the national Swedish newspaper Dagens Nyheter (DN Debatt). This platform is widely used among policy makers and often referred to by other media. For researchers aiming to publish their contributions in DN Debatt, knowing and following media logics in form of the editor's stipulated requirements for submitted texts is decisive. These requirements encompass criteria such as the text's novelty, alignment with the target audience, and whether it presents concrete and conclusive proposals. It is thus important to recognise that the op-ed section does not operate as an unrestricted forum where any issue can be raised for publication; rather, submissions must conform to the medium's editorial standards in terms of content and form. Therefore, the findings cannot be regarded as an absolute evidence of researchers' inclination and capacity to engage in social discourse. Nevertheless, the results shed light on which researcher perspectives receive visibility in the public discourse, the topics they address, and which parts of the public they seek to engage with (Vetenskap & Allmänhet, 2019a). Those issues were addressed in a study analysing researchers' involvement in DN Debatt from 1992 to 2015.

The two research areas that dominated DN Debatt (in pieces addressing issues related to research and science and with one or more researchers as authors) during this period were medicine and social science. Social sciences accounted for two-thirds of the total number of articles, and medicine just under a fifth. The most common subject areas were politics, economics, and health. The results also show that although there have been some shifts over time, DN Debatt remains predominantly a platform for male researchers, typically professors from prominent universities. Thus, the predominant voices shaping public discourse on research in Sweden, as reflected in DN Debatt, are those of middle-aged, white men, who also tend to dominate other mediated public forums such as various news platforms. While the proportion of female authors has gradually increased over time, DN Debatt can be characterised as a relatively static genre/platform. However, changes in the discourse surrounding research may manifest through other media channels, such as social media (Vetenskap & Allmänhet, 2019a).

A study involving a limited cohort of journalists responsible for reporting on research in the Swedish news media (newspapers, radio, and television) points to the complex relationship between journalists and researchers and can also be seen as an illustration of how the media logic works. The selection of research topics by journalists is often guided by their perception of audience interest, as well as by their own and their editors' personal interests in the subject matter. Notably, journalists express a higher level of trust in research within the natural sciences and medicine compared to that within the realms of social science or humanities. Moreover, they tend to prioritise coverage of scientific and medical research due to its perceived ease of acceptance by editors and comprehension by the audience.

Interviews with journalists thus show that natural science and medicine is considered as "true" science, whereas trust in social science and humanities is notably lower (Vetenskap & Allmänhet, 2019b). This is echoed in focus group discussions with members of the public, were participants associate the concept of research predominantly with natural science and medicine, and also expressing a higher level of trust in these domains compared to social sciences and humanities. Research results in science and medicine are perceived as inherently more captivating, valid, and socially beneficial, while those within the domains of social science and humanities are viewed as more abstract, subjective, and less rigorously scientific (Vetenskap & Allmänhet, 2018).

Several journalists express a desire for increased time and resources allocated to investigative science journalism (Vetenskap & Allmänhet, 2019d). They also highlight the challenges for science journalists with the need for thorough review of research findings, something that require both time and expertise. Consequently, journalists often rely on tips from researchers or prominent scientific journals to find potential news stories. However, researchers have expressed concerns that these kinds of relationships can have negative influence on their own careers (Vetenskap & Allmänhet, 2019d).

As been stated above, the issue of trust in relation to science is an important category for understanding contemporary society and it is clear that this trust in several ways is affected mediated communication and the mediatisation process (cf. Rödder & Schäfer, 2010). The next section will present analyses of public trust in research and science in Sweden in general and conclude with a study of public trust and media reports during the Covid pandemic.

Public trust in science

In Sweden, trust in researchers and universities has remained consistently high over the past two decades (see Figure 9.2). Although there is a widespread confidence in research overall, variations appear when examining trust across different research fields. Medicine ranks as the discipline with the highest level of trust, followed by technology, natural sciences, social sciences, educational sciences, and humanities.

We know from previous research that there are differences in levels and forms of trust between different groups of the population and that some decisive factors affect public level of trust in general and also in relation to research and science (Jönsson, 2019; Norén Bretzer, 2017; VA report, 2018:3). Party sympathy is one thing that contributes to explaining the variations in trust that we see among Swedish citizens, and it is above all the sympathisers of the right-wing nationalist party Sweden Democrats who stand out. It has even been stated that there is a polarisation based on party sympathies in the trust in various research disciplines and then especially regarding the view of the humanities. Other aspects that affect level of trust in research and science is the proximity principle (i.e., those who are geographically and culturally close and/or have their own experience of research and universities and colleges) and that highly educated, younger and middle-aged people, as well as those who live in big cities have the highest level of trust in research and researchers (Jönsson, 2019; Jönsson, 2020).

As noted before trust in research is influenced by a number of factors and public's attitudes towards research are largely influenced by the type of research

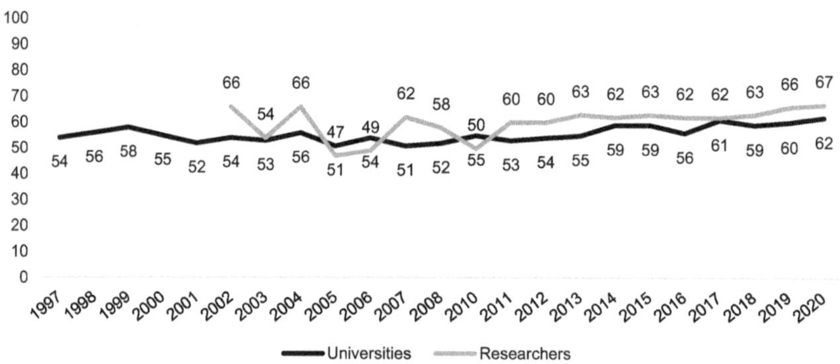

Figure 9.2 Level of trust in universities and researchers among Swedes 1997–2020.

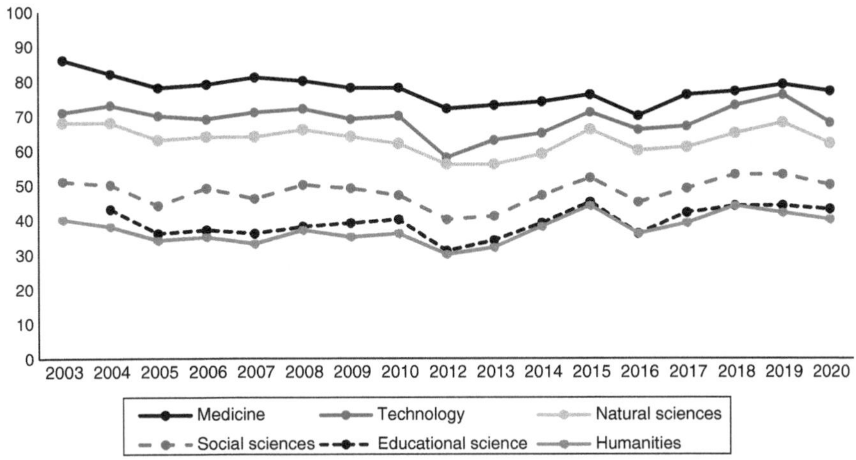

Figure 9.3 Level of trust in different research disciplines among Swedes 2003–2020.

involved (see Figure 9.3). In Sweden citizens have the highest level of trust in the research conducted in the medical field, while research in the humanities and educational sciences recieves the lowest trust (Jönsson, 2021).

The low level of trust in the humanities in relation to other research disciplines in Sweden has sparked some particular interest and research shows that the problem seems to be rather a perceived lack of knowledge about what the humanities really means. A study with the aim to increase understanding about those that claims to have no idea about their confidence in the humanities, among other things shows that two out of three of those who answer "no opinion" of their trust in humanistic research justifies this that they are uncertain on what research within the humanities actually means. If "history" for example is added as an example of research in the humanities, the group that answers "no opinion" decreases and the level of trust increases somewhat. Most trusted is research that is perceived to have a clear social benefit, that it is independent of economic interests and presents results which is confirmed by other research. It is considered less important that the research interests you personally and that it is noticed in the media. However there still is a clear relation between media space and framing and public levels of trust (Vetenskap & Allmänhet, 2022).

To gain deeper insights into what affects levels of trust in research and science, a qualitative study was conducted in 2015. This study involved eight focus groups comprising 45 participants from the general public (Vetenskap & Allmänhet, 2018). The findings from these focus group discussions show that trust in research is strengthened when certain criteria are met. These criteria include:

- The transparency, fairness, comprehensibility, and independence from financial interests in the research process (Process).
- The societal relevance and clarity of research outcomes (Product).

- The involvement of knowledgeable researchers who demonstrate passion for their work (Person).
- The effective communication of research findings in an understandable and engaging manner (Presentation).

Conversely, trust in research diminishes when research is perceived as ambiguous, deceptive, influenced by financial motives, lacking societal benefit, stating obvious findings, presenting erroneous scientific claims, or encountering conflicting research results. Since citizens get most of their information about research and science via media it is thus important that media reports fulfil these criteria. Mediatisation processes as well as an increased professionalisation of science communication is likely to affect trust both in a positive and in a negative way.

Media and trust in a pandemic

The Covid pandemic dominated public discourse on a global scale for several years and also sparked the interest in studies of public trust in science in times of crises (cf. Agley, 2020; Bromme et al., 2022; Petersen et al., 2021; Sulik et al., 2021). During 2020–2021 a study was carried out to investigate how people in Sweden received and interpreted information about the Covid pandemic, and how the pandemic was reported by the media (Vetenskap & Allmänhet, 2020). The main objective of this study was to investigate which factors that influenced people's perceptions in a crisis situation where research and researchers play a central role, in this case during what was also often called an "infodemic" with a constantly changing flow of information. What this study did was to:

1 Monitor the Swedish public's perceptions of the media reporting on the coronavirus and people's confidence in various professions that comment on the virus in the Swedish news media. The collected data consists of 16 online surveys conducted during the period March 2020 – April 2021.
2 Analyse the content of online news articles about the coronavirus and research on Swedish Television (SVT), in the morning daily newspaper Dagens Nyheter (DN) and in the tabloid Aftonbladet at five different time periods between April 2020–May 2021. The time periods coincided with three peaks and two falls in the infection rate and hospital admissions in Sweden.

The results clearly showed that during this period the level of trust in healthcare professionals and researchers when they were visible in Swedish news media speaking about the coronavirus, was high. In fact, there was an increase in the level of trust for all professional groups included in the study, initially in March/April 2020. The rise, which was particularly evident for politicians and officials, however soon went back to previous levels and stabilised (Vetenskap & Allmänhet, 2020). This arched curve is probably due to a so-called "rally-round-the-flag"-effect, which means that the population is rallying behind the country's institutions in times of crisis (Esaiasson et al., 2020).

Among those who say that the pandemic affected their trust levels, a majority say their trust in researchers increased while their trust in journalists decreased. According to the participants their increased trust in researchers is due to the fact that they gained greater insight into how research is carried out and that the pandemic shed further light on researchers work – not least through reporting in the media.

As with public trust in general, there were also during the pandemic differences between different groups in the population. One of the most striking patterns is that those who sympathise with parties to the far right in general have the lowest levels of trust in researchers as well as in journalists. Trust ranking of various news media reporting on the coronavirus is also related to patterns in media use. The public service television company SVT was during the whole period the most widely used news media and also the one with the highest level of trust (Vetenskap & Allmänhet, 2020).

As for the content of the Swedish media news media during the pandemic it was dominated by news about restrictions/guidelines and spread of infection and vaccines. Other common topics were the situation in the hospitals, politics, new research results and economy. The tone of the reporting was mainly neutral, but every fifth article had alarming elements and every ten a more calming note. Researchers (especially from the medical area) were the professional group that was most visible in the news reporting, followed by authority representatives and journalists (Vetenskap & Allmänhet, 2020).

Summary and discussion

In short, when it comes to the question of how research and science is represented in news media, this chapter shows that some disciplines and research issues have a higher news value and are more easily adopted to media logics. Medicine and health issues are two examples of this and although it is also shown that there is generally a high level of trust for science and research (institutions as well as actors) in Sweden, these are also the fields the Swedish public put most trust in and what they think is the most important research. Research in the STEM area is also generally considered more financially profitable and more male dominated compared to social sciences and humanities. On the other hand, we have the humanities that the public indicate they know and trust less that also has less visibility in media discourse. It is concluded that there is a clear link between the media representations and levels of trust in science and research, and that the media discourse on science and research in Sweden is dominated by elite, and to a large extent, male voices.

The higher visibility of certain research areas and topics in the media is affected by different factors. Interviews with journalists indicate that their reporting is influenced by things like their personal interests, what they think will capture audience interest as well as their ideas about what editors would like to see. It is also clear that natural science and medicine is consider as "real" science. Health, for instance, emerges as a clear example of such a topic, and it is also the most common theme in the analysed articles. Social sciences and humanities are on the other hand perceived as more challenging to report on. This perception likely contributes to the dominance of medicine and natural science on science pages of newspapers, while humanities feature prominently on culture pages (Vetenskap & allmänhet, 2019c). While this division may seem natural in relation to the newspaper sections, it potentially shapes societal perceptions of what constitutes research and science.

In the same way focus groups results show that also representatives from the general public mainly associated research with natural science and medicine and that they also trusted these disciplines more compared to social sciences and humanities (Vetenskap & allmänhet, 2018). It is important to note that parts of the difference in trust levels between medicine (ranked highest) and humanities (ranked lowest) is primarily attributed to the percentage of respondents indicating a "lack of understanding." While quite few respondents claim that they are unfamiliar with medicine as a research field, a significant proportion express a lack of understanding about the humanities. Against this backdrop, the newspapers' categorisation of research areas into distinct sections likely shapes the public's perception of research and fortifies a persistent disregard for the humanities as a research domain.

In summary, it is evident that certain research areas and individuals enjoy visibility in the Swedish press, while numerous research findings and researchers remain relatively unnoticed. In order to present a more comprehensive portrayal of research and science and preserve or increase public knowledge of and trust in science, it is imperative for researchers and journalists to engage beyond conventional avenues.

Scientists today encounter numerous challenges in navigating the evolving landscape that questions the role and significance of science. Meeting these challenges necessitates proactive efforts in research communication and enhancing the transparency of research to bolster trust in research and its legitimacy. Iyengar and Massey (2019), for instance, advocate for researchers to develop specialised strategies to counteract various disinformation campaigns, in addition to refining their communication skills.

The level of trust in research and researchers within society is crucial for effectively addressing the challenges that societies face. While maintaining a critical perspective towards elites and public statements is valuable to a certain extent, trust in institutions and individuals remains essential for the functioning of democratic societies. It is important to protect research from being manipulated by political or ideological agendas, especially in the humanities, where trust in research is notably polarised. At the same time it must be noted that the notion of polarisation in itself implies a highlighting of differences, prompting reflection on the potential risks associated with the analytical approaches employed in this chapter and in other contexts. This underscores the need for careful navigation and participation in public discourse on research and science, something that requires more proactive efforts in research communication and an increased transparency. In this "ecology of communication" researchers and scientists need to communicate in several different platforms.

Acknowledgements

The author would like to thank the Swedish organisation *Public & Science* and a special thanks to Dr. Gustav Bohlin as well as Dr. Martin Bergman, Fredrik Brounéus and Dr. Lina Rådmark for cooperation with planning and conducting the studies used in the empirical part of this chapter. The author also wishes to thank the *SOM-institute* in Gothenburg for empirical material about citizen trust.

Notes

1 Science, technology, engineering, and mathematics.
2 Sometimes the concept of medialised is used in parallel.
3 https://vetenskapallmanhet.se/eng/
4 www.gu.se/en/som-institute

References

Abalo, E. (2025). The scientification of risks and the risks of scientification: Insights from the coverage of artificial turf pitches as microplastic pollutants in Sweden. In M.M. Roslyng, A.M. Jönsson, & A. Rantasila, (Eds.), *Communicating science, climate change and the environment in hybrid media: constructed facts, contested truths*. Routledge. Routledge Studies in Media, Communication, and Politics.

Achterberg, P., de Koster, W., & van der Waal (2017). A science confidence gap: education, trust in scientific methods, and trust in scientific institutions in the United States, 2014. *Public Understanding of Science, 26*(6), 704–720. https://doi.org/10.1177/0963662515617367

Agley, J. (2020). Assessing changes in US public trust in science amid the COVID-19 pandemic. *Public Health, 183*, 122–125. https://doi.org/10.1016/j.puhe.2020.05.004

Andersson, U. (2019). Högt förtroende för nyhetsmedier: men under ytan råder stormvarning. High trust in news media: But under the surface a storm is coming'). In U. Andersson,, B. Rönnerstrand, P. Öhberg, & A. Bergström (Eds.), *Storm och stiltje* (Storm and Silence). Gothenburg: SOM Institutet.

Andersson, U., Ohlsson, J., Oscarsson, H., & Oskarson, M. (2017). Larmar och gör sig till. *Andersson, J. Ohlsson, H. Oscarsson & M. Oskarson (red) Larmar och gör sig till. Göteborg: SOM-institutet vid Göteborgs universitet.*

Bauer, M., Ragnarsdottir, A., Rudolfsdottir, A., & Durant, J. (1995). Science and technology in the British press, 1946–1990: a systematic content analysis of the press. In B. Schiele, M. Amyot, & C. Benoit (red.). *When science becomes culture*, Boucherville: University of Ottawa Press

Bauer, M.W., & Bucchi, M. (2007). *Journalism, science and society: between news and public relations*. London, UK: Routledge. https://doi. org/10.4324/9780203942314.

Bauer, M.W., Petkova, K., Boyadjieva, P., & Gornev, G. (2006). Long-term trends in the public representation of science across the 'Iron Curtain': 1946-1995. *Social Studies of Science, 36*(1), 99–131 https://doi.org/10.1177/0306312705053349

Bergström, A. & Oscarsson, H. (2105). Svensk forskningsopinion 2002–2014. In A. Bergström, B. Johansson, H. Oscarsson & M. Oskarsson (red.), *Fragment*. Göteborg: SOM-institutet vid Göteborgs universitet.

Bromme, R., Mede, N.G., Thomm, E., Kremer, B., & Ziegler, R. (2022). An anchor in troubled times: trust in science before and within the COVID-19 pandemic. *PloS One, 17*(2), e0262823.

Bucchi, M., & Mazzolini, R.G. (2003). Big science, little news: science coverage in the Italian daily press, 1946–1997. *Public Understanding of Science, 12*(1), 7–24. https://doi.org/10.1177/09636625030120014 13

Chadwick, A. (2017). *The hybrid media system: politics and power.* Oxford University Press.

Comfort, S.E., Gruszczynski, M., & Browning, N. (2022). Building the science news agenda: the permeability of science journalism to public relations. *Journalism & Mass Communication Quarterly, 101*(3), 637–656. https://doi.org/10.1177/1077699021 1047949

Davies, S.R., & Horst, M. (2016). *Science communication: culture, identity and citizenship.* Springer.

Dudo, A. (2015). Scientists, the media, and the public communication of science. *Sociology Compass, 9*(9), 761–775. https://doi.org/10.1111/soc4.12298

Dunwoody, S. (2021). Science journalism: prospects in the digital age. In M. Bucchi, & B. Trench (Eds.), *Routledge handbook of public communication of science and technology* (pp. 14–32). Routledge.

Esaiasson, P., Sohlberg, J., Ghersetti, M., & Johansson, B. (2021). How the coronavirus crisis affects citizen trust in institutions and in unknown others: Evidence from 'the Swedish experiment'. *European Journal of Political Research, 60*(3), 748–760.

Funk, C. (2017). Mixed messages about public trust in science. *Issues in Science and Technology, 34*(1), 86–88.

Galtung, J., & Ruge, M. (1973). Structuring and selecting news. In S. Cohen, & J. Young (Eds.), *The manufacture of news: Deviance. Social problems and the mass media.* London: Constable.

Gamson, W.A. (1968) *Power and discontent.* Florence, Ky.: Dorsey Press.

Ginosar, A., Zimmerman, I., & Tal, T. (2024). Peripheral science journalism: Scientists and journalists dancing on the same floor. *Journalism Practice, 18*(4), 918–937. https://doi.org/10.1080/17512786.2022.2072368

Goldenberg, M.J. (2023). Public trust in science. *Interdisciplinary Science Reviews, 48*(2), 366–378. doi:10.1080/03080188.2022.2152243

Harcup, T., & O'neill, D. (2017). What is news? News values revisited (again). *Journalism Studies, 18*(12), 1470–1488. https://doi.org/10.1080/1461670X.2016.1150193

Hendriks, F., Kienhues, D., & Bromme, R. (2016). Trust in science and the science of trust. In B. Blo¨baum (Ed.), *Trust and Communication in a Digitized world: Models and Concepts of Trust Research.* https://doi.org/10.1007/978-3-319-28059-2_8

Holmberg, S. & Weibull, L. (2013). Fallet för forskningsförtroendet fortsätter. In L. Weibull, H. Oscarsson, & A. Bergström, (Eds.), *Vägskäl.* Göteborg: SOM-institutet vid Göteborgs universitet.

Iyengar, S., & Massey, D.S. (2019). Scientific communication in a post-truth society. *Proceedings of the National Academy of Sciences, 116*(16), 7656–7661. www.pnas.org/doi/pdf/10.1073/pnas.1805868115

Jönsson, A.M. (2019). Forskarutmaningar (i storm och stiltje). In U. Andersson, B. Rönnerstrand, P. Öhberg & A. Bergström (Eds.), *Storm och stiltje.* Göteborg: SOM-institutet Göteborgs universitet.

Jönsson, A.M. (2020). Polariserat forskningsförtroende – utbildningsnivå och partisympatier vattendelare. In U. Andersson, A. Carlander & P. Öhberg (red) *Regntunga skyar.* Göteborg: SOM-institutet vid Göteborgs universitet.

Jönsson, A.M. (2021). Forskningsförtroende i pandemitider. In U. Andersson, A. Carlander, M. Grusell & P. Öhberg (Eds.), *Ingen anledning till oro (?)*. Göteborg: SOM-institutet, Göteborgs universitet.

Kitcher, P. (2011). Science in a democratic society. In Wenceslao J. Gonzalez (Ed.), *Scientific realism and democratic society*. Leiden, The Netherlands: Brill.

Knudsen, E., Nordø, Å.D., & Iversen, M.H. (2023). How rally-round-the-flag effects shape trust in the news media: evidence from panel waves before and during the covid-19 pandemic crisis. *Political Communication*, *40*(2), 201–221. www.tandfonline.com/doi/pdf/10.1080/10584609.2023.2168322

Mamboleo, G., Chebutuk, R.D., & Matu, N. (2023). Science as a beat in journalism: current status and implications for the future. *Journal of Media and Communication (JMC)*, *2*(1), 66–72.

McCombs, M.E., & Shaw, D.L. (1972). The agenda-setting function of mass media. *Public Opinion Quarterly*, *36*(2), 176–187.

Nelkin, D. (1995). Science controversies: the dynamics of public disputes in the United States. *Handbook of Science and Technology Studies*, *444*, 456.

Nichols, T.M. (2017). *The death of expertise: the campaign against established knowledge and why it matters*. New York, NY: Oxford University Press.

Norén Bretzer, Y. (2017). Förtroendegapet mellan forskningsinstitutioner och forskare. In U. Andersson, J. Ohlsson, H. Ekengren Oscarsson & M. Oskarson (Eds.), *Larmar och gör sig till*. Göteborg: SOM-institutet vid Göteborgs universitet.

Norris, P. (2011). *Democratic deficit: critical citizens revisited*. Cambridge: Cambridge University Press.

Nowotny, H., Scott, P., & Gibbons, M. (2001). *Re-thinking science. Knowledge and the public in an age of uncertainty*, Cambridge: Polity Press.

Olausson, U., & Berglez, P. (2014). Media research on climate change: where have we been and where are we heading?. *Environmental Communication*, *8*(2), 139–141. https://doi.org/10.1080/17524032.2014.910330

Pellechia, M.G. (1997). Trends in science coverage: a content analysis of three US newspapers. *Public Understanding of Science*, *6*(1),49–68.

Peters, B.G. (2002). *Governance: a garbage-can perspective*. Political Science Series 84. Vienna.

Petersen, M.B., Bor, A., Jørgensen, F., & Lindholt, M.F. (2021). Transparent communication about negative features of COVID-19 vaccines decreases acceptance but increases trust. *Proceedings of the National Academy of Sciences*, *118*(29), e2024597118. https://doi.org/10.1073/pnas.2024597118

Roberts, M.R., Reid, G., Schroeder, M., & Norris, S.P. (2013). Causal or spurious? The relationship of knowledge and attitudes to trust in science and technology. *Public Understanding of Science*, *22*(5), 624–641. https://doi.org/10.1177/0963662511420511

Rödder, S., & Schäfer, M. S. (2010). Repercussion and resistance. An empirical study on the interrelation between science and mass media. *Communications*, *35*(3), 249–267.

Rothstein, B. (2005). *Social traps and the problem of trust*. Cambridge: Cambridge University Press.

Schäfer, M. S. (2009). From public understanding to public engagement: an empirical assessment of changes in science coverage. *Science Communication*, *30*(4), 475–505. https://doi.org/10.1177/1075547008326943

Schäfer, M.S. (2011). Sources, characteristics and effects of mass media communication on science: a review of the literature, current trends and areas for future research. *Sociology Compass*, *5*(6), 399–412. https://doi.org/10.1111/j.1751-9020.2011.00373.x

Schäfer, M.S. (2012). Taking stock: a meta-analysis of studies on the media's coverage of science. *Public Understanding of Science*, *21*(6), 650–663. https://doi.org/10.1177/0963662510387559

Schäfer, M.S. (2016). Mediated trust in science: concept, measurement and perspectives for the" science of science communication". *JCOM: Journal of Science Communication*, *15*(5), 1–8.

Schäfer, M.S. (2017). How changing media structures are affecting science news coverage. *The Oxford Handbook of the Science of Science Communication*, *51*, 57.

Stiernstedt, F. (2021). The voices we trust: Public trust in news and information about COVID-19 on Swedish Radio. *Radio Journal: International Studies in Broadcast & Audio Media*, *19*(2), 233–251.

Stosic, M. D., Helwig, S., & Ruben, M. A. (2021). Greater belief in science predicts mask-wearing behavior during COVID-19. *Personality and Individual Differences*, 176, 110769.

Sulik, J., Deroy, O., Dezecache, G., Newson, M., Zhao, Y., El Zein, M., & Tunçgenç, B. (2021). Facing the pandemic with trust in science. *Humanities and Social Sciences Communications*, *8*(1), 1–10. https://doi.org/10.1057/s41599-021-00982-9

Summ, A., & Volpers, A. M. (2016). What's science? Where's science? Science journalism in German print media. *Public Understanding of Science*, *25*(7), 775–790. https://doi.org/10.1177/0963662515583419

Turska-Kawa, A., Csanyi, P., & Kucharčí, R. (2022). From the" rally'round the flag" effect to a social crisis of confidence: Poland and Slovakia in the first year of the covid-19 pandemic.

Vetenskap & Allmänhet (2011). *Slitsamt förvärvat och snabbt fördärvat?* (VA-rapport 2011:1). Stockholm: Vetenskap & Allmänhet.

Vetenskap & Allmänhet (2014). *Fusk och förtroende – om mediers forskningsrapportering och förtroendet för forskning*, VA-rapport 2014:3.Stockholm: Vetenskap & Allmänhet.

Vetenskap & Allmänhet (2018). *Vad påverkar förtroende för forskning? – Fokusgrupper med allmänheten.* VA-rapport 2018:5. Stockholm: Vetenskap & Allmänhet.

Vetenskap & Allmänhet (2019a). *Lika barn debattera mest – forskares medverkan på DN-Debatt 1992-2015.* (Rapport 2019:2). Stockholm: Vetenskap & Allmänhet.

Vetenskap & Allmänhet (2019b). *Bakom rubrikerna – intervjuer med 12 journalister om att bevaka forskning* (Rapport 2019:3). Stockholm: Vetenskap & Allmänhet.

Vetenskap & Allmänhet (2019c). *Forskning i svensk press 1995–2015 – en innehållsanalys* (Rapport 2019:7). Stockholm: Vetenskap & Allmänhet.

Vetenskap & Allmänhet (2019d). *Jag vill, men hinner inte! Forskares syn på kommunikation och öppen vetenskap.* VA-rapport 2019:8. Stockholm: Vetenskap & Allmänhet.

Vetenskap & Allmänhet (2020). Corona i svenska medier – ökat förtroende för myndighetspersoner. Hämtad från https://v-a.se/2020/04/corona-i-svenska-medier-okat-fortroende-for-myndighetspersoner/

Vetenskap & Allmänhet (2022) *Varför så dålig hum om humaniora? – Förtroende för humanistisk forskning i Sverige.* VA-rapport 2022:4. Stockholm: Vetenskap & Allmänhet.

Weingart, P. (2022). Trust or attention? Medialization of science revisited. *Public Understanding of Science*, *31*(3), 288–296. https://doi.org/10.1177/09636625211070888

Weingart, P., Engels, A., & Pansegrau, P. (2000). Risks of communication: discourses on climate change in science, politics, and the mass media. *Public Understanding of Science*, *9*(3), 261–284

Weingart, P., & Guenther, L. (2016). Science communication and the issue of trust. *Journal of Science communication*, *15*(5), C01.

Weitkamp, E., Milani, E., Ridgway, A., & Wilkinson, C. (2021). Exploring the digital media ecology: insights from a study of healthy diets and climate change communication on digital and social media. *Journal of Science Communication*, *20*(3), A02.

10 Constructing trust through affective discipline

Finnish nuclear energy experts and the Fukushima Daiichi disaster

Anna Rantasila

Introduction

Techno-scientific phenomena, such as public health, technological accidents and natural disasters, are rife with affect and emotion (Oikkonen, 2017), and the meltdown at the Fukushima Daiichi Nuclear Power Plant in Japan in March 2011 was no exception. According to Oikkonen (2017), scholarly inquiries informed by affect provide crucial new angles to understand how representations of techno-scientific phenomena are constituted. Thus, understanding the role of affect and emotion in disruptive events such as the 2011 Fukushima Daiichi disaster can offer insight for both media and expert organisations regarding how to better respond to the public's need for information.

In public discussions held in the late 2010s, concern was expressed over the role of scientific expertise and the burgeoning lack of trust in established scientists, as climate change denial and other conspiratorial beliefs circulated in various online spaces (e.g. Capstick & Pidgeon, 2014; Forchtner & Lubarda, 2022). However, the COVID-19 pandemic and its socio-political and cultural aftermath brought these concerns to the forefront to a far great extent. Thus, I believe that revisiting an older case concerning a disruptive event, such as the Fukushima Daiichi disaster in 2011, which spotlighted questions about the credibility and trustworthiness of scientific and technological experts, can provide some insight into how trust in the scientific establishment is constructed in the mainstream news media. At a time when scientific knowledge and informed analysis have been increasingly challenged by conspiratorial claims and "alternative facts" of various strands circulated in the hybrid media environment (see, e.g. Demata, 2025; Larsen & Roslyng, 2025; Mäkinen, 2025, this volume), successful communication about crises and risks requires an understanding of the affective underpinnings of techno-scientific phenomena and how they come to be represented in mainstream media as a part of a hybrid media environment (Chadwick, 2017; Sumiala et al., 2018).

In this chapter, I address the construction of trust by analysing how the Radiation and Nuclear Safety Authority (STUK; *Säteilyturvakeskus* in Finnish) in Finland and the Finnish public service broadcasting company *Yleisradio* (YLE) engaged in maintaining and rebuilding trust in scientific and technological expertise during the 2011 Fukushima Daiichi nuclear disaster. I study the construction of trust through

DOI: 10.4324/9781003479550-13

the concept of *affective discipline* (Laaksonen & Rantasila, 2021; Rantasila, 2020, 2022), which refers to a set of dynamics and practices aimed at managing public emotions (Rantasila, 2022). As trust can be understood as an emotion (Belli & Broncano, 2017), I suggest that the notion of affective discipline can be used to discuss how the media appearances of experts during times of crisis attempt to manage public emotions and trust in scientific and technological institutions.

The chapter aims to answer the following research questions:

1 What kind of language did STUK representatives use in their appearances in YLE online news stories, and did they or journalists directly mention or address emotions in these stories?
2 How was the notion of trust constructed in the Fukushima coverage?
3 Did the construction of trust involve acts of affective discipline?

Based on my previous analysis of interviews and other media appearances of officials from the STUK in YLE's online coverage of the Fukushima Daiichi nuclear disaster between 2011 and 2016 (Rantasila, 2018; Valaskivi et al., 2019b), I suggest that during the acute phase of the disaster in March 2011, mainstream news media and experts from the STUK were implementing communication strategies that were simultaneously aimed at informing the public and *managing the public emotions*. In the news coverage of the Fukushima Daiichi disaster, this often resulted in assigning the public a dual role: the stories addressed their readership as a rational public, while the public was simultaneously represented in the stories as misled and panicky (Valaskivi et al., 2019a). Furthermore, according to my previous study of online comments on news reports about Fukushima Daiichi (Rantasila, 2018; Valaskivi et al., 2019b), this strategy often resulted in mixed reactions in the readership and occasionally appeared to reinforce existing prejudices towards scientific expert organisations as uncaring and dismissive.

I now discuss the empirical context of this chapter by providing an overview of the events leading up to the meltdown of three reactors at the now-disabled Fukushima Daiichi Nuclear Power Plant in March 2011. Thereafter, I provide an overview of nuclear energy in Finland, followed by a presentation of the key theoretical elements of the study at hand. I then elaborate on the concept of affective discipline and discus why notions of affect and emotion are crucial in crisis communication and journalism and why they matter for the topic of this book. I then present the empirical material and analysis method of the study. Finally, I present the results of the analysis and discuss the theoretical and practical implications of the study results.

Empirical context: Japan and Finland

Affect and emotion are tied to collective, or intersubjective, historical, social, cultural and discursive circumstances where feelings and sensations take place (Ahmed, 2004; Wetherell, 2015). Moreover, feelings and sensations are tied to the culturally, socially and historically shared systems and processes of meaning-making

(Wetherell, 2015). Therefore, to understand the Finnish news coverage of the Fukushima Daiichi disaster and the positions of experts, I provide a socio-political background of nuclear energy in the Japanese and Finnish contexts to support both the empirical material of this chapter and my analysis.

The triple disaster of March 2011 in Japan

On March 11, 2011, 14:46 JST, a magnitude 9.0 to 9.1 earthquake occurred in the Pacific Ocean approximately 70 km east of the Oshika peninsula and 440 km north-east of Tokyo at a depth of circa 30 km. The earthquake was the strongest ever recorded in Japan and fourth strongest in the world; it was so powerful that it moved the whole island of Honshu more than 2 km eastward and shifted the Earth's axis (Lochbaum et al., 2014). It set off a tsunami wave that travelled at a speed of 700 km/h, in some places reaching almost 10 km inland. According to Japan's Reconstruction Agency (2023a), 19,765 people lost their lives in the natural disaster. The combination of a record-breaking earthquake and a tsunami would have been devastating on their own, but the north-eastern coast of Japan was also the home of one of the largest nuclear power stations in the world, Fukushima Daiichi.

The now-disabled Fukushima Daiichi Nuclear Power Plant is located roughly 260 km north of Tokyo between the towns of Okuma and Futaba on the Pacific coast of the Fukushima prefecture. In operation since 1971, the plant is operated by Japan's largest electric utility Tokyo Electric Power Company (TEPCO). It has six boiling water reactors, three of which (Units 4–6) were out of commission for maintenance on March 11, 2011. When the earthquake struck the plant, all of its emergency measures initially worked as designed. The plant also had a seawall, which deflected the first tsunami wave. Unfortunately, the second tsunami wave was 14 m tall, exceeding the height of the seawall by 4 m (Kurokawa et al., 2012a; Lochbaum et al., 2014).

The tsunami wave that reached the plant destroyed the emergency seawater pumps meant to carry excess heat from the reactors and flooded the basements of most of the buildings on the site, effectively obliterating the emergency diesel generators meant to power the plant's cooling systems. The water also damaged the electrical distribution systems and backup batteries; thus, the power plant had neither external nor internal power supply. The surging water also spread debris throughout the power plant site, making it more difficult for workers to move into the area to begin repairs once the waters had retreated. On March 11, 2011, 19:00 JST, Prime Minister Naoto Kan declared a nuclear emergency (Kurokawa et al., 2012a; Lochbaum et al., 2014).

As the reactors at Fukushima Daiichi were not being cooled, their temperature began to rise, and the cooling water began to evaporate, leaving the nuclear fuel rods inside the reactor exposed. Because of the power outage, plant workers had no data on the situation, but they were aware of the risks and tried to do what they could in a very precarious situation. On March 12, 2011, hydrogen inside the Unit 1 reactor building exploded. By the evening of March 12, the fuel in Units 1 and 2 was in meltdown, and Unit 3 was also having serious cooling issues. On March 14,

Unit 3 exploded, and a third explosion occurred on March 15, when Unit 4 blew up. The explosions in Units 3 and 4 were also caused by a build-up of hydrogen gas. Moreover, the efforts to bring the situation under control were hampered by continuous earthquake aftershocks, inclement weather, electric cables that were too short, the fact that the earthquake and tsunami had damaged roads and infrastructure around the struggling power plant, etc. (Kurokawa et al., 2012a, 2012b; Lochbaum et al., 2014).

Altogether, approximately 164,000 people were evacuated from the area surrounding Fukushima Daiichi between March and April 2011, and 167 workers were exposed to high doses of radiation (Kurokawa et al., 2012b; Reconstruction Agency, 2017). The nuclear emissions contaminated 1800 square kilometres of the area around the plant, and in 2012, the cumulative radiation dose per adult was five micro sieverts or higher annually, significantly higher than the average of 3.75 micro sieverts for an adult in Japan per year (Kurokawa et al., 2012b; Reconstruction Agency, 2016). However, by November 2023, the evacuation orders were lifted from most areas (Reconstruction Agency, 2023a). Nevertheless, as of June 2024, some towns and villages remained classified as "difficult-to-return-to" by the Reconstruction Agency (2023a), and the annual cumulative radiation dose per adult also returned to the average levels for Fukushima prefecture (Reconstruction Agency, 2023b).

As several studies and the official Japanese investigation commissions into the disaster have stated, the shortcomings at Fukushima Daiichi were not just a series of unexpected and unavoidable accidents; rather, many of the problems at Fukushima Daiichi, from neglected safety measures to incompetent company management, had deep roots in the operational culture of the Japanese nuclear industry (Government of Japan, 2011, 2012; Hatamura et al., 2012; Kurokawa et al., 2012a, 2012b; Lochbaum et al., 2014). While the meltdowns may not have been entirely preventable, the extent of the damage could have been significantly lower had the plant operator TEPCO and the Japanese officials responsible for implementing and overseeing nuclear energy policy taken several previous warning signals more seriously (Kurokawa et al., 2012b; Lochbaum et al., 2014).

Nuclear power in Japan has been tied to economic growth and the provision of heavy industries, with reasonably priced electricity in the post-war reconstruction effort (Yoshimi & Loh, 2012). The interconnectedness of industry and politics in the nuclear energy sector in Japan has been dubbed a "nuclear village", denoting the close-knit relationships and vested interests of the actors involved in the production, consumption and security of nuclear energy. The nuclear village has upheld a "myth of safety" around Japanese nuclear energy, according to which nuclear energy is safe because of the superior know-how and standards of Japanese science and engineering, whereby major accidents are highly unlikely (Nöggerath et al., 2011; Penney, 2012; Suzuki, 2011; see also Kajser et al., 2021; Walker, 2010). This led to a mindset in which financial profits trumped safety, and addressing safety issues meant additional costs and the admission of potential failure (Yoshimi & Loh, 2012). As the investigations into the aftermath of the 2011 disaster have revealed, TEPCO deliberately ignored several warnings by both Japanese and international inspectors at Fukushima Daiichi for several years before the disaster (Kurokawa et al., 2012b; Lochbaum et al., 2014; Penney, 2012).

Nuclear energy in Finland

The Finnish nuclear energy programme was initiated in 1955, and the national regulatory official, the STUK, was established in 1958 to regulate radiation equipment in hospitals. In the late 1960s, the STUK was appointed by the Ministry of Social Affairs and Health as an independent regulator to oversee the safety of the emerging nuclear power industry. There are currently two nuclear power plants in Finland, one located in Olkiluoto in Southwestern Finland and the other in Loviisa in Southeastern Finland. The Olkiluoto plant has three reactor units: two built during the 1970s and 1980s and the third completed in 2022. The Loviisa plant has two reactors, both built in the 1970s.

In the post-Second World War period, between the 1950s and 1980s, nuclear energy was tied to rapid industrialisation in Finland. According to studies and opinion polls conducted by universities, nuclear energy advocates and media organisations, a small majority of Finns had been consistently pro-nuclear. However, the 1986 Chernobyl disaster did cause, at least temporarily, an observable decline in those in favour of nuclear energy (Timonen et al., 1987; see also Gamson & Modigliani, 1989). Moreover, the Chernobyl disaster made the public suspicious of the STUK because of poor communication with the public and media as well as complicated and contradictory advice provided during the accident and the ensuing fallout (Timonen et al., 1987). Following the Chernobyl accident, the national debate about nuclear energy in Finland concerned the processing and deposition of nuclear waste between 1995 and 2001 (Raittila, 2000, 2001). The construction of the Onkalo nuclear waste repository began in 2004, and the topic has not been discussed critically in the mainstream media ever since.

The Fukushima Daiichi disaster in March 2011 coincided with a parliamentary election campaign in Finland, and the safety of nuclear energy became a minor campaign issue (Rantasila, 2020). Moreover, the STUK was again under fire after their website crashed temporarily on March 11 and 12 when thousands of users tried to access the site for information about Fukushima Daiichi (Rantasila, 2020). As with Chernobyl, the Fukushima Daiichi accident led to an increase in the number of anti-nuclear responses in opinion polls, but the numbers have since rebounded.

As in Japan, nuclear energy policy in Finland has been entwined with questions around economic growth and the success of the nation's heavy industries. Another major theme has been energy independence, particularly after Russia's invasion of Ukraine in 2022. In fact, this invasion led to the cancellation of a power plant project with Russia's Rosatom in Northern Finland. However, despite sporadic opposition, the image of the nuclear industry is less mired in scandal in Finland than in the United States, South Korea or Japan, for example, and is perhaps viewed more positively than in other European countries (Laihonen, 2016; Ruostetsaari, 2018; Vehkalahti, 2017; see also Abe, 2013; Jasanoff & Kim, 2009; Kim et al., 2013; Penney, 2012; Weart, 2012). Because of the relatively positive image, the risks of nuclear power or nuclear waste are rarely discussed in the Finnish mainstream news media. The 1986 Chernobyl disaster, discussions of the nuclear waste repository in the 1990s and the 2011 Fukushima Daiichi disaster are the most recent exceptions. In the case of Fukushima Daiichi, the mediated public debate was often framed in such a way that removed the notion of risk from nuclear energy in Finland and placed it on nuclear energy elsewhere, that is, in Japan (Valaskivi et al., 2019a). While this claim is technically accurate given the varying geophysical conditions, reactor types and safety features, such statements portray an attitude that resembles the above-mentioned Japanese myth of safety around nuclear energy.

Theoretical premises and related work

In this section, I discuss the theoretical premises of the chapter and elaborate on how the notions of affect and affective discipline can be used to analyse the construction of trust in scientific expertise during a disruptive event. In addition, I explore the notion of trust as an emotion and reflect on theoretical findings in the context of trust research and science communication. While trust research mainly focuses on business and administration studies, some themes do overlap with those in science communication, science and technology and media and journalism studies and provide fruitful insights that can be combined with theories of affect.

Affect and affective discipline

Since the early 2000s, the concept of affect has gained traction in various fields, including media and journalism and science and technology studies (e.g. Oikkonen, 2017; Papacharissi, 2014; Wahl-Jorgensen, 2019, 2020; Wetherell, 2015). In my previous work (Rantasila, 2018, 2020; Valaskivi et al., 2019a, 2019b), I attempted to approach affect as a complex set of intensities and sensations that are set into motion in encounters between bodies (broadly defined) and events in specific socio-cultural and historical contexts. Following Paasonen (2015) and Oikkonen (2017), I suggest that affect is a biological, subjective and autonomous response, one strongly connected to culturally, socially and historically circulated emotions (Ahmed, 2004; Massumi, 1995). However, at the level of lived experience and empirical reality, it is difficult to distinguish affect and emotion, as they are entangled in a continuous loop of reactions, recollections and reiterations (Ahmed, 2004; Paasonen, 2015; Oikkonen, 2017; Wahl-Jorgensen, 2019; Wetherell, 2015).

In this chapter, I examine the notion of affect in the context of mediated representations of the Fukushima Daiichi nuclear disaster and interviews with scientific and technological experts. Traditionally, institutions of journalism and science have sought to distance themselves from emotions as a way of bolstering their notion of objectivity (Pantti, 2010; Wahl-Jorgensen, 2019, 2020). However, the concept of affect renders visible the complex relationship with news media, techno-scientific expertise, crisis communication and the public, as it brings attention to, for example, what kinds of emotions are named, discussed and managed and which ones remain undefined and vague (Valaskivi et al., 2019a). In this chapter, I follow

Ahmed's (2004) notion that certain affects tend to become attached to certain bodies, and I focus on how affect and emotion were articulated by experts and how these experts stuck to the public image of the Fukushima Daiichi disaster itself.

Trust as emotion

Moreover, affect is most intimately related to the topic of this book, that of constructing trust. Trust can be defined in various ways. *The Oxford English Dictionary* (2015) defines it as a "Firm belief in the reliability, truth, or ability of someone or something; confidence or faith[…]". In social psychology, trust also requires the notion of risk and, thus, an element of vulnerability (Blomqvist, 1997). In the context of science communication, Hendriks and Kienhues (2019) understand trust as defined by a dependent relationship between a trustor and a trustee, which is based on positive expectations. According to their example, in science communication, laypeople must assess the trustworthiness of scientific experts before trusting their claims.

Furthermore, Belli and Broncano (2017) categorise trust as a "meta-emotion" that precedes other emotions. According to them, meta-emotions are related to other emotions and connect emotions to each other; for instance, love between romantic partners requires trusting each other. However, Belli and Broncano argue that trust (like all emotions) is highly contextual, noting that it is a different thing to trust a bus driver than to trust your spouse. Partly because of this fluidity, Belli and Broncano argue that trust is rarely recognised as an emotion as such. They further maintain that meta-emotions such as trust are crucial in managing and regulating other emotions: a breach of trust in a relationship can shift a feeling of friendship into anger and betrayal, but it can be mended by building more trust. While the authors' examples are from interpersonal and intimate relationships, I suggest that these elements can be applied to examining trust in scientific expertise and affective discipline in mediated communication.

Constructing trust through affective discipline

The concept of affective labour (Hardt, 1999) can be used to discuss situations involving the construction of feelings, such as security and trust, in emotional management in contexts such as war and terror (Anderson, 2010). Zhukova's (2015, 2016) work on the 1986 Chernobyl nuclear disaster and its aftermath demonstrates the trust-building efforts of scientists and officials in Belarus and Ukraine regarding the risks posed by the disaster. In addition to the notion of affective labour, I suggest that the emerging concept of *affective discipline* can be used as a lens to examine the affective labour of trust-building by journalists and scientific experts as an attempt to manage the affect and emotions of the general public during disruptive events, such as the Fukushima Daiichi nuclear disaster or the COVID-19 pandemic (Laaksonen & Rantasila, 2021; Rantasila, 2022).

By affective discipline, I refer to instances in which *an institution or individual engages in practices that seek to sustain or renew the affective and emotional tone of a given situation* (Laaksonen & Rantasila, 2021; Rantasila, 2018, 2022). These instances are often related to situations where an event has disrupted the everyday routine. In my previous work (Rantasila, 2022; Laaksonen & Rantasila, 2021), I examined how participants in online discussions seek to return the tone of the discussion to the status quo after another participant has disrupted it with a comment that has been deemed inappropriate by others. For example, in a social media discussion about Finns who had been unaccounted for in Japan in March 2011, commenters expressed their concerns about the missing people until one person pointedly asked, "Who cares about a couple of Finns abroad when we have hundreds of people dying as a result of domestic violence and alcohol here [in Finland]" (Rantasila, 2022)? This comment effectively turned the thread into a discussion about what was wrong with the comment and commenter and why it was inappropriate.

While I have previously applied the notion of affective discipline to social media discussions, I suggest that the concept can be a fruitful way to examine how news media and scientific experts engage in acts of (or at least attempts at) managing the affective reactions and expressed emotions of the public during disruptive events. As Horsley (2016) points out, the news media play a significant role in defining when a situation is or is not a disaster and who is involved in it (Rogers & Pearce, 2016). Moreover, as Wahl-Jorgensen (2019) notes, emotion is used strategically in journalism, which requires an understanding of the rules governing public displays of emotion. Therefore, I suggest that the management of publicly displayed emotions by journalists and scientific experts in journalistic narratives is crucial for constructing trust: which providers are deemed trustworthy of information by the mainstream media, and how is this trustworthiness constructed in the news coverage?

Materials and methods

The empirical material for this study was collected as a part of a larger project examining media coverage of the Fukushima Daiichi disaster between 2014 and 2016 (Valasikivi et al., 2019a, 2019b), and some of the data have been analysed in publications that were part of my doctoral dissertation (Rantasila, 2018, 2020, 2022).

The empirical data consist of online news stories about the Fukushima Daiichi disaster published by the YLE. The news stories were collected from YLE's online Finnish-language news portal *yle.fi/uutiset* (yle.fi/news) between 2014 and 2016, and the sample consisted of 554 online news items published between March 11, 2011, and December 31, 2016. The news stories were retrieved from YLE's website through the site's search function using the keywords "Fukushima*", "Japan*", and "säteily*" (Finnish for "radiation") and by running corresponding searches on Google. The stories were saved as copies to the Zotero software and on a separate secure hard drive by the author. For detailed analysis, the articles were converted into.rtf files and analysed in the NVivo software.

To examine the construction of trust in news media during the most acute phase of the Fukushima Daiichi disaster as well as narrow down the sample, I searched the stories from 2011 ($N = 304$) either for direct quotations or mentions of actors who were defined as experts (Valaskivi et al., 2019b). This yielded a sample of 127 stories, 49 of which had direct quotations from 43 individuals. The use of experts by YLE in their Fukushima Daiichi coverage was highly domestic in nature: of the 43 people cited, 30 were Finnish, 11 of whom were from the STUK. Six other STUK experts were quoted indirectly, bringing the total number of individual STUK experts to 17. Two of the STUK representatives, the director general Jukka Laaksonen and a senior inspector Riku Mattila, were quoted in eight stories, while most STUK experts appeared in a single story each. Altogether, the 17 STUK representatives were featured in 32 stories, which are discussed in greater detail below. Of all the experts (the STUK and others) quoted in these 32 stories, four were female, and 19 were male. Also noteworthy is that all 32 stories were published between March 11 and 31, 2011, when the nuclear disaster was unfolding in the aftermath of the earthquake and tsunami.

Drawing from the theoretical premises presented above, I analyse the stories qualitatively. I use the methods of close reading, metaphor analysis and qualitative content analysis to understand how emotions and trust were constructed in the news stories and whether the STUK experts engaged in acts of affective discipline. Following Lakoff and Johnson's (1981) approach to metaphor analysis, I identify the uses of metaphors in each article and examine the contexts where they were used. I also look at how emotion was connected to these metaphors. I then proceed to examine the ways in which the metaphors were connected to the broader context of the stories as well as how emotions were presented in the stories – if at all.

The stories were coded manually through a hermeneutic coding approach, where I first coded the stories as I read them closely, identifying elements related to the three aspects mentioned above: metaphor, affect and emotion and trust. This initial round of coding resulted in 50 individual codes. In the next phase, I grouped these initial codes into seven themes. The themes and number of references are presented below in Table 10.1 below.

Following this, I discuss the findings in more detail by first examining the use of metaphors in the stories in general. I then delve deeper into the role of the experts

Table 10.1 Themes and number of references in the sample of 32 news stories

Theme	*Number of references*
Uncertainty, risk & threat	277
Trust	113
Experts and organisations	112
Emotion	86
Metaphor	69
Health effects	56
Other nuclear disasters	23

in the stories, explore how the events at Fukushima Daiichi and their impact are discussed and how emotion, trust and affective discipline are present in the sample.

Findings

Generally speaking, YLE's news coverage of the Fukushima Daiichi disaster in March 2011 was heavily technology-oriented, and the 32 stories that featured STUK experts were concerned with specific issues, particularly the probability of a meltdown and the amount of radiation released from the damaged reactors. The focus of the stories was either on events at the Fukushima Daiichi site in Japan or the possible impact of the radiation leakage on Finland and Finns. The Japanese people living in the affected area were mentioned only in passing.

The language deployed in the stories, both in the quotations and journalistic narrative, was relatively neutral but rife with numbers and specific terms mostly related to the measurement of radiation doses and exposure limits. In the sample of 32 stories, the use of metaphors was mostly limited to describing the damaged reactors through words related to conflict and battle (Lakoff & Johnson, 1981), while the rescue efforts in the plant area were referred to as "a struggle" or "a fight" either against time or natural elements. In a few instances, the radiation leak was referred to as "an escape", and the spraying of water into the damaged reactor buildings was called "shooting". The reactor itself was referenced using bodily metaphors, such as describing the fuel rods inside the damaged nuclear reactors as the "heart" or "core" of the reactor or discussing the release of steam from the reactor buildings as "blowing" or "venting".

Uncertainty and risk

Much more common than the use of metaphors were references to uncertainty, risk and threat. As Table 10.1 indicates, this was the most common theme in the stories. However, it was also the most diverse in terms of the types of phrases it contained, as there were both direct and indirect ways of expressing uncertainty in the Finnish language. While the metaphors were mostly present in the journalistic narrative of the events, this category included phrases containing direct quotations from both the experts and the journalistic narrative.

For example, in a story from March 13 (YLE News, 2011a) recounting the latest information about the damaged reactors, uncertainty was expressed in the head-note of the story as follows: "However, the STUK *has no exact information* about the fuel damage or *possible* core meltdown in reactor three of the [Fukushima Daiichi] plant".[1] In the second paragraph of the story, an STUK official was quoted as saying, "According to the information we received from the Swedish officials, water is fed into reactor three at the moment and the heart of the reactor is under-water. *However, we don't have any official information* on *whether* the heart [of the reactor] went dry overnight *or not*".

The above example illustrates a pattern that plays out throughout the sample, where both the journalistic narrative and the experts featured in the article appeared very cautious about saying anything for certain. The two excerpts above also highlight another prominent feature of the sample: uncertainty over the events was often paired with either a sense of risk or threat, such as core meltdown and fuel damage or the risk of radiation leaks or exposure.

Trust and distrust

Phrases and sentences coded as related to trust and distrust were also among some of the most common themes, with 113 references. As with the phrases related to uncertainty, trust and distrust often appeared together in the same story and, thus, are discussed here as a single theme. For instance, in stories published on March 16 and 27, 2011, STUK Director General Jukka Laaksonen expressed great distrust towards Japanese nuclear officials while simultaneously highlighting the expertise and excellence of Finnish nuclear scientists and industry operators. In the opening paragraph of the March 16 story (YLE News, 2011c), Laaksonen was quoted as saying, "*Together with our colleagues from the IAEA, we've been perplexed about how hard it seems* [for the Japanese] to deal with the rescue efforts. *Apparently, they don't have capable local management* there to run the effort". Similarly, the headnote of the March 27 story (YLE News, 2011e) summarised the story as follows: "The *Finnish Radiation and Nuclear Safety Authority* is *concerned about the contradicting and confusing nature of Japan's information* on its damaged plants. *Director General Jukka Laaksonen estimates that the STUK experts have the best insight* on the situation in Japan".

In other words, the above stories construct a two-sided narrative whereby the Japanese nuclear experts were presented as unreliable and incapable, while the Finns were reliable, if not the best. As noted in my previous work (Rantasila, 2018), the irony of this narrative did not escape social media commenters, and Laaksonen's remarks drew public backlash, particularly after the story published on March 16. This also reflects Hendrik and Kienhues's (2019) and Belli and Broncano's (2017) notions about trust as a two-way relationship.

Furthermore, the discussions around trust and distrust focused on the trustworthiness of nuclear technology or numerical data, such as radiation measurements. In addition, natural elements related to the disaster area, such as the Pacific Ocean or Japan's geographic location in relation to Finland, were discussed as factors relating to uncertainty or certainty. For example, several stories emphasised that the safety mechanisms at Fukushima Daiichi functioned as designed until the second tsunami wave breached the seawall and flooded the site.

Emotion and acts of affective discipline

In terms of phrases and words coded as referring to emotions, there were altogether 86 references in the sample. Among these references, there were 54 instances of a specific emotion, such as fear or concern. Somewhat surprisingly, 32 instances referred to optimistic assessments of the situation facing Fukushima Daiichi or Finland, either explicitly citing hopefulness or implicitly narrating the developments at the stricken power plant as promising. However, as with trust and distrust, positive and negative emotions often appeared simultaneously in the stories, constructing an either/or type of narrative where some events were described as worrisome while some invoked hope for a good outcome. The emotional narrative was also tied to the construction of trust in these stories.

For example, in a story published on March 23, 2011, about the first indicators of radioactive material from Fukushima Daiichi detected in Finland (YLE News, 2011d), the STUK officials noted that the amounts detected were so small that they could only be identified with special laboratory equipment. "These amounts are *absolutely nothing to worry about*", an official was quoted as saying. However, the story then continued by stating that "the STUK *has been following the radiation levels in Finland more closely* during the past week. With the methods used, even extremely small changes in the radiation levels can be detected". In other words, while the radiation situation in Finland was understood as sufficiently serious to be monitored, as officials remained vigilant, there was no need for the general public to worry about it. The above example also illustrates a subtle way of engaging in affective discipline: the news media and officials featured in the coverage constructed what can be understood as proper ways (for Finns) of reacting to the situation at Fukushima Daiichi. The public was told not to worry, implying that outward expressions of concern by laypeople would be inappropriate.

A more direct example of affective discipline was observed in a story published March 14, 2011, titled "Physics Professor: No Reason for Nuclear Power Hysteria" (YLE News, 2011b). The story was a relatively long interview with a professor emeritus of physics. It mostly revolved around the professor's insights on the reliability of information about the disaster and what the members of the public should think about the situation. The story began with a quotation from the professor stating that "Currently, there is a lot of *emotional coverage* on the news". It then continued with the professor critiquing German media for negative and anti-nuclear coverage. However, the overall tone of the quotations from the professor was that of reassurance, even to the point of arrogance, for example, that Finns had nothing to worry about, that the STUK had accurate and verified information, that local power plants were as safe as they could be, that there would be no "new Chernobyl" and "that a fallout from Fukushima would spread to Finland is *totally impossible*". In other words, the desired affective response to the disaster was that of calm, collective and rational behaviour, similarly to the previous example.

Discussion

As Ahmed (2004) and Wetherell (2015) note, affect sticks and accumulates over time. From the perspective of the construction of trust and acts of affective discipline, the STUK expert interviews in YLE's coverage of Fukushima Daiichi presents an interesting example of how the accumulation of affect appears as a key element in acts of affective discipline employed to construct trust and reassurance in scientific expertise. The construction of trust via acts of affective discipline appeared to rely heavily on the presentation of scientifically accurate facts and figures, as the nuclear industry expert interviews in YLE's coverage of Fukushima Daiichi focused on experts providing technical information to the public about the situation in Fukushima while also reiterating that there was no reason for concern in Finland (see also Valaskivi et al., 2019b; Rantasila, 2018). Thus, the stories were characterised by tension between the journalistic narrative and getting the scientific facts right, often resulting in a great deal of technical descriptions dominating the coverage.

In some stories, such as the interviews with the STUK Director General Jukka Laaksonen and the physics professor, the acts of affective discipline were almost banal in the way in which the experts told the public how to behave. However, in most stories, the strategic use of emotion (Wahl-Jorgensen, 2019) in building emotional tension in the stories – by constructing contrasts between certainty and uncertainty, trust and distrust and hope and concern – was key in nudging the public's emotions. One very practical reason behind the interplay of certainty and uncertainty may have been the fact that the information about the events *was* indeed uncertain at the Fukushima Daiichi site as the disaster unfolded. Moreover, while journalists tended to prefer clear-cut facts, scientists tended to be painfully aware that very few things were certain. This unwillingness of the STUK experts to provide bold statements about the situation at Fukushima Daiichi may have had the unintended consequence of constructing an unnecessary level of uncertainty to the journalistic narrative.

In addition to the tension between trust and distrust, YLE's coverage of Fukushima Daiichi also points to how news media may simultaneously acknowledge and dismiss public emotions (Valaskivi et al., 2019a, 2019b). However, this dynamic often becomes visible only when the stories are examined together as a cumulative narrative instead of individually. While the events in Fukushima Daiichi were described in some stories as worrisome by both journalists and experts, other stories in the same sample asked the same implied audience not to worry, as Finland was safe. This accumulation of contradictory messages may explain why a certain level of distrust towards both mainstream news media and scientific experts has appeared to persist from crisis to crisis, as illustrated by, for example, the chapters of this book exploring the COVID-19 pandemic.

Conclusion

The role of scientific expertise in YLE's coverage of the Fukushima Daiichi disaster in 2011 provides an interesting case to examine how trust as an emotion is constructed and managed in the news media. On one hand, both journalists and scientific experts appeared to be invested in getting the facts right and providing trustworthy information to the public, thus contributing to their sense of safety. On the other hand, the uncertainty of the unfolding disaster constructed another layer of uncertainty that appeared difficult to avoid. As Rogers and Pearce (2016) note, successful risk and crisis communication requires an approach that considers not only quantitative, fact-based knowledge dissemination but also one that acknowledges the emotional responses behind risk perception. However, as both the analysis of the news stories above and my previous work (Rantasila, 2018, 2022; Valaskivi et al., 2019b) suggest, merely stating that "there is no reason for concern" does very little to assuage fear and suspicion. Moreover, as the Fukushima Daiichi disaster illustrates, public perceptions of risk and disaster are entangled with affective intensities that stem from the cultural and social history of previous disasters (Ahmed, 2004; Jasanoff & Kim, 2009; Valaskivi et al., 2019a; Weart, 2012). In addition, as Sarlos argues in the following chapter of this volume, news media also may have a tendency to amplify risk perception related to nuclear energy.

The construction of trust in scientific expertise thus requires knowledge of techno-scientific facts as well as an understanding of how human emotions are tied to decision-making systems. As examples of more recent disaster and crisis coverage in this volume illustrate, making this connection can be a fruitful way of tackling anti-science sentiments that may have detrimental effects on individuals and the planet.

Acknowledgements

The empirical data used in this chapter were collected through a project funded by the Academy of Finland (grant number #277465) and the Japan Society for the Promotion of Science (grant number #14544647), "Media Events, Circulation and Emerging Social Media Practices: Tracing the Meaning of Fukushima", which ran from 2014 to 2016. The other members of the project provided their consent to the use of the data in this chapter.

Note

1 All quotations translated by the author.

References

YLE News stories:

YLE News. (2011a). STUK: Vaurioituneiden reaktoreiden tilanne vakiintumassa [STUK: Damaged reactors stabilising]. Published March 13, 2011, at 08:04. Available at https://yle.fi/a/3-5324305.

YLE News. (2011b). Fysiikan professori: Ydinvoimahysteriaan ei syytä [Physics professor: No reason for nuclear power hysteria]. Published March 14, 2011, at 16:59. Available at https://yle.fi/a/3-5324999.

YLE News. (2011c). STUKin pääjohtaja arvostelee kovin sanoin Japanin pelastustoimia [STUK director general harshly criticizes rescue efforts in Japan] Published March 16, 2011, at 12:18. Available at https://yle.fi/a/3-5326479.

YLE News. (2011d). Helsingissä havaittu Fukushiman radioaktiivisia aineita [Radioactive substances from Fukushima detected in Helsinki]. Published March 23, 2011, at 15:10. Available at https://yle.fi/a/3-5330400.

YLE News. (2011e). STUKin Laaksonen: Meillä on ehkä paras kuva Japanin tilanteesta [STUK's Laaksonen: We probably have the best insight on Japan]. Published March 27, 2011, at 11:21. Available at https://yle.fi/a/3-5332677.

Peer-reviewed literature

Abe, Y. (2013). Risk assessment of nuclear power by Japanese newspapers following the Chernobyl nuclear disaster. *International Journal of Communication*, *7*(22), 1968–1989. https://ijoc.org/index.php/ijoc/article/view/1848

Ahmed, S. (2004). *Cultural politics of emotion*. Edinburgh University Press.

Anderson, B. (2010). Modulating the excess of affect. Morale in a state of "total war." In M. Gregg & G.J. Seigworth (Eds.), *The affect theory reader* (pp. 161–185.). Duke University Press.

Belli, S., & Broncano, F. (2017). Trust as meta-emotion. *Metaphilosophy*, *48*(4), 430–448. https://doi.org/10.1111/meta.12255

Blomqvist, K.-M. (1997). The many faces of trust. *Scandinavian Journal of Management*, *13*(3), 271–286. https://doi.org/10.1016/S0956-5221(97)84644-1

Capstick, S.B., & Pidgeon, N.F. (2014). What is climate change scepticism? Examination of the concept using a mixed methods study of the UK public. *Global Environmental Change*, *24*, 389–401. https://doi.org/10.1016/j.gloenvcha.2013.08.012

Chadwick, A. (2017) The hybrid media system: politics and power (2nd ed.). Oxford University Press. https://doi.org/10.1093/oso/9780190696726.001.0001

Demata, M. (2025). Fighting (for) truth? Alex Jones, the WHO and the legitimation of conspiracy discourse. In M.M. Roslyng, A. Rantasila, & A.M. Jönsson (Eds.), *Communicating science, climate change and the environment in hybrid media: Constructed facts, contested truths* (pp. 89–111). Routledge.

Forchtner, B., & Lubarda, B. (2022). Scepticisms and beyond? A comprehensive portrait of climate change communication by the far right in the European Parliament. *Environmental Politics*, *32*(1), 43–68. https://doi-org.ezproxy.cc.lut.fi/10.1080/09644016.2022.2048556

Gamson, W.A., & Modigliani, A. (1989). Media discourse and public opinion on nuclear power: a constructionist approach. *American Journal of Sociology*, *95*(1), 1–37. https://doi.org/10.1086/229213

Government of Japan. (2012). *Convention on nuclear safety national report of Japan for the second extraordinary meeting*. Available at: https://fdada.info/en/home2/link2/library2/. Accessed October 15, 2020.

Government of Japan, Nuclear Emergency Response Headquarters. (2011). *Additional report of the Japanese Government to the IAEA. The accident at TEPCO's Fukushima Nuclear Power Stations* (2nd report). Available at: www.iaea.org/sites/default/files/japanreport120911.pdf. Accessed February 12, 2025.

Hardt, M. (1999). Affective labor. *Boundary*, *26*(2), 89–100.

Hatamura, Y., Oike, K., Kakinuma, S., Takasu, Y., Takano, T., Tanaka, Y., Hayashi, Y., Furukawa, M., Yanagida, K., & Yoshioka, H. (2012). *Final report: investigation committee on the accident at Fukushima nuclear power stations of Tokyo electric power company*. Available at: https://fdada.info/en/home2/link2/library2/. Accessed October 15, 2020.

Hendriks, F., & Kienhues, D. (2019). Science understanding between scientific literacy and trust: contributions from psychological and educational research. In A. Leßmöllmann, M. Dascal, & T. Gloning (Eds.), *Science Communication* (pp. 29–50). De Gruyter Mouton. https://doi.org/10.1515/9783110255522-002

Horsley, S.J. (2016). Media framing of disasters: implications for disaster response communicators. In A. Schwarz, M.W. Seeger, & C. Auer (Eds.), *The handbook of international crisis communication research* (pp. 230–243). John Wiley & Sons. https://doi.org/10.1002/9781118516812.ch15

Jasanoff, S., & Kim, S.-H. (2009). Containing the atom: sociotechnical imaginaries and nuclear power in the United States and South Korea. *Minerva, 47*, 119–146. https://doi.org/10.1007/s11024-009-9124-4

Kaijser, A., Lehtonen, M., Meyer, J.H., & Rubio-Varas, M. (2021). *Engaging the atom: the history of nuclear energy and society in Europe from the 1950s to the present.* West Virginia University Press.

Kim, Y., Kim, M., & Kim, W. (2013). Effect of the Fukushima nuclear disaster on global public acceptance of nuclear energy. *Energy Policy*, *61*, 822–828. http://doi.org/10.1016/j.enpol.2013.06.107

Kurokawa, K., Ishibashi, K., Oshima, K., Sakiyama, H., Sakurai, M., Tanaka, K., Tanaka, M., Nomura, S., Hachisuka, R., & Yokoyama, Y. (2012a). *The official report of the Fukushima nuclear accident independent investigation commission.* An executive summary. The National Diet of Japan. Available via an archived website of the commission at: http://warp.da.ndl.go.jp/info:ndljp/pid/3856371/naiic.go.jp/en/. Accessed October 25, 2019.

Kurokawa, K., Ishibashi, K., Oshima, K., Sakiyama, H., Sakurai, M., Tanaka, K., Tanaka, M., Nomura, S., Hachisuka, R., & Yokoyama, Y. (2012b). *The official report of the Fukushima nuclear accident independent investigation commission.* The National Diet of Japan. Available via an archived website of the commission at: http://warp.da.ndl.go.jp/info:ndljp/pid/3856371/naiic.go.jp/en/. Accessed October 25, 2019.

Laaksonen, S.M. & Rantasila, A. (2021) Rocketing sheep: affective discipline in anonymous mobile social media Jodel during the Covid-19 pandemic. *AoIR Selected Papers of Internet Research.* https://doi.org/10.5210/spir.v2021i0.12198.

Laihonen, M. (2016). Political foreplay for nuclear new build. Defining good at the intersection of politics, economy and technology [Doctoral dissertation, Aalto University]. Aalto University Publication Archive. https://aaltodoc.aalto.fi/handle/123456789/24604

Lakoff, G., & Johnson, M. (1981.) *Metaphors we live by*. University of Chicago Press.

Larsen, G. & Roslyng, M.M. (2025). Knowledge and counter-knowledge. The construction of facts in the vaccination debate. In M.M. Roslyng, A. Rantasila, & A.M. Jönsson (Eds.), *Communicating science, climate change and the environment in hybrid media: Constructed facts, contested truths* (pp. 112–130). Routledge.

Lochbaum, D., Lyman, E., & Stranahan, S.Q. (2014). *Fukushima. The story of a nuclear disaster*. The New Press.

Massumi, B. (1995). The autonomy of affect. *Cultural Critique*, *31*, 83–109. https://doi.org/10.2307/1354446

Mäkinen, M. (2025). Citizen activists or pandemic deniers? Alternative voices in the Finnish journalistic media during the COVID-19 pandemic. In M.M. Roslyng, A. Rantasila, & A.M. Jönsson (Eds.), *Communicating science, climate change and the environment in hybrid media: Constructed facts, contested truths* (pp. 131–150). Routledge.

Nöggerath, J., Geller, R.J., & Gusiakov, V.K. (2011). Fukushima: the myth of safety, the reality of geoscience. *Bulletin of the Atomic Scientists*, *67*(5), 37–46. https://doi.org/10.1177/0096340211421607

Oikkonen, V. (2017). Affect, technoscience and textual analysis: interrogating the affective dynamics of the zika epidemic through media texts. *Social Studies of Science*, *47*(5), 681–702. https://doi.org/10.1177/0306312717723760

Oxford English Dictionary (2015). Trust. https://doi.org/10.1093/OED/5777528687

Paasonen, S. (2015). A midsummer's bonfire: affective intensities of online debate. In K. Hillis, S. Paasonen, & M. Petit (Eds.), *Networked* affect (pp. 27–42). MIT Press. https://doi.org/10.7551/mitpress/9715.001.0001

Pantti, M. (2010). The value of emotion: an examination of television journalists' notions on emotionality. *European Journal of Communication*, *25*(2), 168–181. https://doi.org/10.1177/0267323110363653

Papacharissi, Z. (2014). *Affective publics: sentiment, technology, and politics*. Oxford University Press. https://doi.org/10.1093/acprof:oso/9780199999736.001.0001

Penney, M. (2012). Nuclear nationalism and Fukushima. *The Asia Pacific Journal Japan Focus*, *10*(11), 1–21. https://apjjf.org/2012/10/11/matthew-penney/3712/article. Accessed May 31, 2024.

Raittila, P. (2000). *Ydinjätteen loppusijoitus mediassa. Julkishallinnon ydinjätetutkimusohjelmaan liittyvä joukkoviestinnän seurantatutkimus* [*Permanent storage of nuclear waste in the media. A mass media monitoring study for the public government's nuclear waste study program*]. Tampere University Press.

Raittila, P. (Ed.). (2001). *Mediat ydinjätettä hautaamassa. Eri intressiryhmien julkisuuteen pääsy, dialogi ja argumentointi ydinjätteen loppusijoitusta koskevassa keskustelussa 1999–2001* [The media burying the nuclear waste. Access to publicity by different interest groups, their dialogue, and argumentation in discussion about the permanent storage of nuclear waste in 1999–2001]. Tampere University Press.

Rantasila, A. (2018) Tahmaiset affektit. Fukushima Daiichin ydinonnettomuus YLE:n uutisoinnin verkkokommenteissa [Sticky affects. Fukushima Daiichi nuclear accident in YLE News' online comments]. *Lähikuva*, *31*(3), 30–45. https://doi.org/10.23994/lk.76570

Rantasila, A. (2020). *Circulating emotions, sticky feelings. Affective dynamics of the Fukushima Daiichi nuclear disaster in a hybrid media environment* [Doctoral dissertation, Tampere University]. Tampere University Press. http://urn.fi/URN:ISBN:978-952-03-1764-5

Rantasila, A. (2022). Managing unpleasant moods: affective discipline in Facebook discussions. *European Journal of Cultural Studies*, *25*(1), 256–274. https://doi.org/10.1177/1367549420973217

Reconstruction Agency, (2016). *Basic information on radiation risk*. Available at: https://www.google.com/url?sa=t&rct=j&q=&esrc=s&source=web&cd=1&cad=rja&uact=8&ved=2ahUKEwjAnfadqrflAhUryaYKHbZYDeUQFjAAegQIABAC&url=https%3A%2F%2Fwww.reconstruction.go.jp%2Fenglish%2Ftopics%2FRR%2F20160308BasicInformationRRen.pdf&usg=AOvVaw1spn2uXJ1BUcyPopV7s4yb.

Reconstruction Agency. (2017). *The process and prospects of reconstruction*. Available at: http://www.reconstruction.go.jp/english/topics/Progress_to_date/index.html.

Reconstruction Agency. (2023a). *Current status of reconstruction and future efforts*. Available at https://www.reconstruction.go.jp/english/topics/Progress_to_date/English_December_2023_genjoutorikumi-E.pdf.

Reconstruction Agency. (2023b). *Eliminating negative reputation impact. Reconstruction from nuclear disaster & the history, safety and revitalization of Fukushima*. Available at https://www.reconstruction.go.jp/topics/main-cat1/sub-cat1-4/fuhyou/pamphlet/latest/huhyou-higai-husshoku_E.pdf.

Rogers, M.B., & Pearce, J.M. (2016). The psychology of crisis communication. In A. Schwarz, M.W. Seeger, & C. Auer (Eds.), *The handbook of international crisis communication research* (pp. 76–89). John Wiley & Sons. https://doi.org/10.1002/9781118516812

Ruostetsaari, I. (2018). Citizens' support for rival modes of political involvement in energy policymaking. The case of Finland in 2007–2016. *International Journal of Energy Sector Management, 12*(3), 323–340. https://doi.org/10.1108/IJESM-12-2017-0005

Sumiala J., Valaskivi, K., Tikka, M.M.T., & Huhtamäki, J. (2018). *Hybrid media events: The Charlie Hebdo attacks and the global circulation of terrorist violence*. Emerald.

Suzuki, T. (2011). Deconstructing the zero-risk mindset: the lessons and future responsibilities for a post-Fukushima nuclear Japan. *Bulletin of the Atomic Scientists*, *67*(5), 9–18.

Timonen, I., Kallio, R., & Mörä, T. (1987). *Tshernobylin voimalaonnettomuus lehdistössä* [Chernobyl nuclear power plant accident in the press]. Tampere University Press.

Valaskivi, K., Rantasila, A., Tanaka, M., & Kunelius, R. (2019a). The global circulation of affect: the case of Iodide tablets. In K. Valaskivi, A. Rantasila, M. Tanaka, & R. Kunelius (2019). *Traces of Fukushima. Global events, networked media and circulating emotions* (pp. 101–118). Palgrave Macmillan.

Valaskivi, K., Rantasila, A., Tanaka, M., & Kunelius, R. (2019b). Affective entanglements of expertise: the Finnish case. In K. Valaskivi, A. Rantasila, M. Tanaka, & R. Kunelius. *Traces of Fukushima. Global events, networked media and circulating emotions* (pp. 119–136). Palgrave Macmillan.

Vehkalahti, P. (2017) *Pohjoisen ydinmylly. Julkinen keskustelu Fennovoiman ydinvoimalasta 2007–2013* [Nuclear furnace of the north: Public discussion about Fennovoima's nuclear power plant from 2007 to 2013] [Doctoral dissertation, University of Tampere]. Tampere university Press. http://urn.fi/URN:ISBN:978-952-03-0541-3

Wahl-Jorgensen, K. (2019) *Emotions, media and politics. Contemporary political communication*. Polity.

Wahl-Jorgensen, K. (2020). An emotional turn in journalism studies? *Digital Journalism*, *8*(2), 175–194. https://doi.org/10.1080/21670811.2019.1697626

Walker, B.L. (2010). *Toxic archipelago: a history of industrial disease in Japan*. University of Washington Press.

Weart, S.R. (2012). *The rise of nuclear fear*. Harvard University Press.

Wetherell, M. (2015). Trends in the turn to affect: a social psychological critique. *Body & Society*, *21*(2), 139–166. https://doi.org/10.1177/1357034X14539020

Yoshimi, S., & Loh, S.-L. (2012). Radioactive rain and the American umbrella. *The Journal of Asian Studies*, *71*(2), 319–331.

Zhukova, E. (2015). *Whose responsibility? Chernobyl as trauma management in Belarus and Ukraine* [Doctoral dissertation, Aarhus University]. Forlaget Politica.

Zhukova, E. (2016). From ontological security to cultural trauma: the case of Chernobyl in Belarus and Ukraine. *Acta Sociologica*, *59*(4), 332–346. https://doi.org/10.1177/0001699316658697

11 Nuclear stories in the news media

Filtering and altering of expert views

Gabor Sarlos

Introduction

Media channels play an essential role in presenting nuclear power both as a long-term, safe and low-cost form of clean energy production, and emphasising the risks related to uranium mining, long-term nuclear waste management and human errors in planning and operating nuclear power plants (Culley et al., 2010). Supporters claim that nuclear energy is indispensable in the fight against global warming and climate change by providing a carbon-free form of large-scale energy source. Opponents claim that nuclear energy production is not safe and low-cost at all, and point out that media representation of its environmental impact should be based on a full life cycle analysis, including uranium mining and nuclear waste management, as well (Egres & Sarlos, 2023). Gamson and Modigliani (1989) claim that media plays an essential role in framing nuclear power as a technological advancement. Perceptions of nuclear accidents are also influenced by the position the media takes: a pro-industry, pro-regulator approach alienates risks while an anti-nuclear position reinforces the relevance of taking negative incidents into account (Angelique & Cunningham, 2006). The media portrayal of risks becomes a decisive factor in formulating public opinion about nuclear energy (Kristiansen, 2017a). Opponents claim warily that the nuclear industry has always been strong in radiating a positive vision about the industry but has had difficulties in handling the challenges of the present (Sarlos, 2014). Contrasting views demonstrate that differences in views relate to questions about ownership of knowledge and what the "right" level, source and type of knowledge is.

Perception of nuclear energy has changed significantly over the years. Starting with the 1950s, nuclear power was first seen as a source of endless energy, upon which industry and household electricity needs could be safely built. By the early 1970s, it reached the status of the key depository of economic wealth, and a symbol of social and technological progress (Van Der Pligt, 1985). The same decade saw a growing level of concern and criticism towards nuclear energy, partly because of the type of hazards it presented, and partly because of the political ripples the anti-nuclear movement caused in Europe and North America (Fischhoff et al., 1978, 1981). The perception of atomic power was dramatically marred by the Chernobyl (Chornobyl in Ukrainian) nuclear accident in 1986. It then eventually regained

DOI: 10.4324/9781003479550-14

its place and status, leading to its consolidation in the energy mix of 50 countries around the world (IAEA, 2024a). The Fukushima accident in 2011 reinforced concerns about the safety of nuclear power, while the increasing awareness of the climate crisis led to the repositioning of nuclear energy as the only reliable, stable source of clean energy (Pidgeon et al., 2008; Katz-Rosene, 2021). Today, nuclear power enjoys a mixed perception (Slovic et al., 2000). Occasional operational failures and accidents, identified through their geographical names such as Sellafield, Three Miles Islands, Chernobyl, and Fukushima, appear to leave a long-term imprint on public perception of nuclear power (Sarlos & Fekete, 2019). Risks are counterbalanced by the benefits of portraying nuclear energy as an essential, carbon-free alternative to fossil fuels. Influenced by the complexity of the technology and the diversity of contexts and framings, consequently, differences in views about the related risks and benefits prevail (Vainio et al., 2017). Media plays an essential role in interpreting and contextualising the risk factors of nuclear energy production. Together with individual value systems and worldviews (Beck, 1999), media will continue shaping public perception of nuclear energy in the future.

Media coverage of nuclear energy, specifically nuclear accidents have been researched extensively (Kinsella, 2005; Doyle, 2011; Hara, 2013; Rieu, 2013). There has been less focus however on analysing the role of nuclear energy industry actors on the presentation of nuclear-related content in key news media. Traditionally, communication activities of companies, organisations and associations in the field are discussed in the context of organisational communication, while media coverage is analysed in the media frame and discourse analysis, and media models related to risk and crisis communication (Bakir, 2010; Kristiansen, 2017b; Vossen, 2020).

The current chapter discusses the role news media plays in framing and interpreting nuclear-related and nuclear organisations-originated information by using the framework of risk communication to connect the areas of organisational communication and media communication. The study intends to contribute to the discourse on how reality is constructed through news media frames in a field considered technologically complex and identified with significant risk factors. It adapts the Social Amplification of Risk (SARF) conceptual framework and identifies news media as "social amplification stations" in the communication process. It also covers the importance of the actual communicator, as the effectiveness and impact of risk communication depends to a large extent on the credibility of the disseminator of the risk message.

Nuclear organisations invest significant energy into influencing the agenda-setting and framing activities of media, to have their interpretation of developments represented in media news. The current study analyses the impact the risk-related press releases of 2023 of the International Atomic Energy Agency (IAEA) have on the nuclear-related news coverage of four leading international news media outlets, Al-Jazeera, Euronews, Politico and Reuters. The IAEA is an autonomous international organisation within the United Nations and the leading international organisation in the field of nuclear energy, its interpretation of nuclear risks is considered authoritative. The study investigates to what extent the four media channels share the risk perception views of the IAEA and, interpreting it from the perspective of the public, what the nuclear risks are that it filters or alters in its news communication.

Research objective and question

The research aims to understand in what form news media acts as social amplification stations in disseminating nuclear risk messages. According to the Social Amplification of Risk Framework, media can exercise a dual role; it can amplify risk messages through its filtering function and attenuate risk messages through its altering function. The research aims to identify patterns of these dual functions in the case of nuclear-related media communication. The analysis furthermore intends to identify whether the amplification or attenuation of risk messages is influenced by the acceptance of the original disseminator of the message.

The research question is therefore the following:

> ***In news media coverage of nuclear energy, is amplification and attenuation of risk messages influenced by the credibility of the original source of the information?***

Conceptual context

The Social Amplification of Risk Framework (SARF)

The Social Amplification of Risk Framework focuses on identifying causes and mechanisms of distinguishing between risks identified as critical by experts and risks felt strongly by the public. It studies the process of hazards interacting with the various processes and value sets of the individual that lead to the amplification or the attenuation of public responses to risk. The concept underlines that public recognition of specific risks is heavily influenced by socio-group behaviours, individual value systems and filters, information processes and institutional structures. One of the critically important elements in this risk perception process is that of media (Kasperson et al., 1988).

The SARF conceptual framework grants a specific status to media by acknowledging its essential role in conveying information. Media is identified as one of the "social amplification stations" in the risk awareness processes. "Social amplification stations generate and transmit information via communications channels. In addition, each recipient also engages in amplification and attenuation processes, thereby acting as an amplification station for risk-related information". (Kasperson et al., 1988, p.181). The ultimate interpretation of the signal, conveyed by the various social amplification stations, still lies with the individual recipient of the message.

Kasperson et al. (1988, 2022) furthermore claim that amplification or attenuation occurs at two distinct stages: in the transfer of information to the public about the risk and the response mechanism of society. The authors furthermore claim that "risk events interact with psychological, social, and cultural processes in ways that can heighten or attenuate public perceptions of risk and related risk behavior" (Kasperson et al., 1988, p.178.). Arvai and Rivers (2013) claim that the technical aspect of risk dominates the discourse and the social, political and cultural dynamics that influence risk management and risk perception are underappreciated. In their view, risk communication is ineffective if it focuses on providing additional technical information only. They acknowledge the ongoing influence of the "knowledge deficit model", which, in their view, should build on a broader spectrum of communication, acknowledging the social, cultural and value set differences (Renn, 2010).

Due to previous accidents, the nuclear industry is confronted with a level of "stigma". Stigma, in the SARF context, refers to a negative image identified with undesirable individual, social groups, activities or organisations (Flynn, 2003). Negative associations with the nuclear industry most frequently occur in connection with the perceived mismanagement and human errors related to the TMI, Chernobyl and Fukushima disasters. In the case of the most recent accident, the Fukushima disaster, this stigma leads to scepticism and wariness towards the Japanese government, authorities and TEPCO, the company operating the Fukushima Daiichi nuclear plant (de Groot et al., 2013; Kusumi et al., 2017). Similarly, stigma was left on the Soviet authorities at the time of the Chernobyl accident (Miller, 2016). Stigma carries a spatial and temporal element as well; the closer the accident is in terms of time and proximity, the bigger the stigma might be. However, Goatcher and Brunsden (2011) claim that in the case of major nuclear accidents, there are practically no precise spatial or temporal boundaries. Names of earlier accidents act as triggers where the imprint is embedded in the collective mindset for a long time. The extent and the long-term effects of the catastrophe can lead to a collective, or even cultural trauma of the society, inducing fundamental reluctance and opposition to the use of nuclear energy (Assmann, 2008; Novikau, 2017; Rantasila et al., 2018; Zhukova, 2016).

Following the Fukushima nuclear disaster in 2011, the social amplification of perceived risks influenced nuclear policies in several countries, including for example Japan, Germany, and Belgium (Hagen et al., 2022). Kristiansen et al. (2016) however view a gradually growing acceptance of nuclear energy, following a sharp drop immediately after the Fukushima disaster. Visschers and Siegrist (2013) believe that negative associations and loss of trust influence perceptions negatively for a limited time only and claim that the public has a relatively stable attitude towards nuclear power.

The last two decades have seen the digital transformation of media, with the social media revolution empowering every actor to gain quasi-direct information on any given topic and become active makers of media themselves. Still, traditional media continues to have an essential function in educating, entertaining and informing the public. Both traditional and social media play an inherent role in shaping public perception of nuclear energy, including the weight of associated risks and benefits (Weart, 2012).

By coding, decoding and granting social value to signals, media actively "filters and alters" signals. The two functions might take place in two separate stages. In the first stage, the media filters the topics it intends to cover, consequently fulfilling its agenda-setting function (McCombs & Shaw, 1972). The second stage includes the alteration of the original risk message and leads to its amplification or attenuation, as interpreted in the SARF framework.

In the latter stage, media performs the important step of framing, by creating the elements of the perceived reality it finds important or relevant. Framing helps "make texts more salient in such a way as to promote a particular problem definition, causal interpretation, moral evaluation, and / or treatment recommendation for the item described" (Entman, 1993). Framing therefore is used in this research to identify and articulate patterns that are embedded into the text of each unit. These frames exercise a significant influence on how the reader interprets a given text, as they help set the text into a more precise understanding. By referring to structures familiar to the reader, these frames provide the link between one's existing knowledge and values, and the new pieces of information from the text.

Within the SARF framework, media can both amplify and attenuate risk perceptions and consequently increase or decrease public recognition of risks. In its original format, the SARF framework described the impact of media stations as "ripples", sharing information in widening concentric circles. It was compared to ripples in a pond to identify with impacts created within the social amplification context, with risks spreading outward or staying inward (Pidgeon et al., 2013).

However, due to the rapid emergence of social media, the interpretation of ripples lost its validity, while media continues to play an essential role as a social amplification station.

Methodology

To provide an analysis of the texts and provide answers to the research questions, a series of research tools were used. The scope of the research had to be clearly defined, including the choice of organisation as original source of information, selection of media as the "social amplification stations" and the definition of the period of the study.

The choice of organisation and focus on IAEA, the International Atomic Energy Agency is justified by acknowledging the role the organisation plays as "the global focal point for nuclear cooperation" (IAEA, 2024a). In this capacity IAEA "develops nuclear safety standards and, based on these standards, promotes the achievement and maintenance of high levels of safety in applications of nuclear energy, as well as the protection of human health and the environment against ionizing radiation" (IAEA, 2024b). There are several other international organisations in the field of nuclear energy, but the IAEA is the only truly global and not vendor-driven organisation for nuclear energy. It is furthermore the only global organisation in the field of nuclear energy that regularly issues a significant number of press releases. To analyse these through both quantitative and qualitative methods, a minimum sample size was considered important.

Table 11.1 Key data of the four news media companies

	Ownership	*HQ*	*Nature*	*Focus*
Al-Jazeera	Qatar Media Company	Qatar	multimedia	Middle East, Asia, North Africa
Euronews	individual investors	Brussels, Belgium	multimedia	Europe
Politico	Axel Springer SE	Virginia, USA	online newspaper	Europe, North America
Reuters	Thomson Reuters Corporation	London, UK	news agency	global

The choice of media was driven by the intention to have an overview of leading English-language news media on a global scale. The four media show significant differences in ownership, origins, type of media and operational models, however, their diversity was one of the key considerations in involving them in the research. In selecting the four media for study, global relevance and representation of the diversity of the world were considered essential. The decision was made for media that have a politically aware but generalist approach and that would consider their role as interpreting global developments to a wide and aware public. In practical terms, it was furthermore important to be able to access their database of English-language articles without the need for additional payment. Finally, their databases needed to offer searchability by both the abbreviation and the full name of the IAEA, the International Atomic Energy Agency (Table 11.1).

The decision to choose a full year as the time frame of the study was based on the need to have an appropriate period of time for the identification of key risk patterns without the risk of one single event dominating the analysis. Setting identical timeframes for the analysis of the IAEA press releases of IAEA and the media coverage of the four media channels served to analyse closely the direct risk filtering and altering function of the media. In practical terms, it was important to see what immediate impact the IAEA press releases made on the choice and interpretation of nuclear-related news published by the media outlets. Finally, preliminary research showed that the diversity and abundance of nuclear-related news in 2023 made it an appropriate year for study.

Frame analysis was used to define essential risk claims in the selected leading, international English-language news media. Texts were interpreted through both qualitative and quantitative approaches to identify risk amplification patterns. The analysis was based on matching the 157 press releases issued in 2023 by the IAEA, with the 256 English-language news media articles of Al-Jazeera, Euronews, Politico and Reuters, concerning the IAEA in the same 12 months period. Each press release and news media article formed a unit of the study.

Texts were read closely and analysed by two experienced coders. Coders defined the key patterns of the units following a pilot version of coding, with the coding process being controlled on a randomly designed basis. The use of AI-driven Chat

GPT 4 was applied, to systematically extract and condense key information from the texts, enhancing the efficiency of the content analysis. Through the use of artificial intelligence, the probability of identifying recurring patterns and trends was increased. The information extracted and condensed through AI was then manually checked and confirmed by the coders. The coded data was then used to identify risk amplification frames.

Standardisation of coding was ensured by focusing on text with a clear risk component. Through the coding, first the type of nuclear risk was identified, followed by coding by the country in focus, the key message, the strength of the risk component, the role of IAEA in the risk communication and the communication patterns the units contained. Textual analysis of each press release covered identification of the key message, speaker, topic category and risk category. Analysis of the media articles looked at the key message, role of IAEA in the article, speaker (or speakers) in the article, topic category, risk category, weight of risk as well as the use of any visual material. By analysing the IAEA press releases and the media articles separately, two separate data sets were created. Comparison of these data sets then allowed to identify the role media channels played in nuclear-related news and define the impact the credibility of IAEA can play in the amplification or attenuation of risk messages.

Findings

The first stage of the analysis identified 157 IAEA press releases and 256 articles with reference to IAEA in 2023. Of the 157 press releases, 72 referred to Ukraine and 85 covered a further 38 countries, with Japan represented frequently with 16 press releases. A similar relative dominance of Ukraine appeared in the articles: 107 of the 256 articles referred to nuclear-related developments in Ukraine, while 149 articles covered other countries. Within the "international" articles, Iran and Japan were represented significantly, with 46 and 41 articles each. Due to the relative dominance of Ukraine in both press releases and articles and to avoid overrepresentation of one single aspect, the war in Ukraine, it was decided that the topic of Ukraine be discussed separately from the other topics and countries. By having two separate geographical foci of discussion, it was envisaged that the chance for identifying specific characteristics can be increased. This way the relative dominance of the war in Ukraine and its impact on nuclear power coverage does not overshadow other important trends and frames. For ease of understanding, the two groups of press releases and news articles were grouped according to geographical categories, "Ukraine" and "International". Choosing these categories aligns closely with the activities of IAEA which focuses on providing expertise to individual countries, adapted to the level of their nuclear development needs.

In both the Ukraine and the International sections, the identification of key frames is followed by a short textual analysis and one or more citations from the original press releases or media articles to illustrate the pattern or risk communication.

Ukraine

In 2023, 107 articles appeared in the news media about Ukraine with the mention of IAEA, shared between the media outlets as follows:

- Al-Jazeera: 9 articles
- Euronews: 21 articles
- Politico: 12 articles
- Reuters: 65 articles

In the same period, the relatively biggest number of the 72 press releases about Ukraine referred to primarily to its nuclear infrastructure and the electricity supply, followed by military developments and water and cooling issues.

The focus of media interest differed from the key focus of IAEA press releases. Media articles covered primarily the military developments impacting nuclear infrastructure, with a focus on the effect on the Zaporizhzhia nuclear power plant. This was followed by the topic of reactor cooling, mostly related to the breach of the Kakhovka dam. Finally, two key IAEA initiatives, the creation of a safety zone and the approval of both parties of the five safety principles were also covered extensively in the news media. (Table 11.2).

The difference in the interest of IAEA and the media is illustrated by the difference in their choices of topics. The relative media interest in each press release is evidenced by the article/press release ratio. The two press releases about the IAEA initiative of five principles and the safety zone were met with the biggest interest, leading to 13 articles and a 6.5 article/press release ratio. On the other end of the spectrum, while the 18 IAEA press releases covered the topic of infrastructure and electricity supply extensively, these generated only three articles, leading to a 0.16 article/press release ratio. The filtering function of the media is a representation of the editorial interests, in comparison of the communication interests of the IAEA.

Table 11.2 Key risk areas in the news media articles and IAEA press releases about Ukraine

	Number of articles – Ukraine	*Number of IAEA press releases – Ukraine*	*Article/press release*
TOTAL, of which:	**107**	**72**	**1.48**
Military	30	13	2.3
Water and cooling	18	13	1.38
Infrastructure and electricity supply	3	18	0.16
IAEA initiative of five principles and safety zone	13	2	6.5
Other:	43	26	

Risk communication frames about Ukraine

Sixty four out of one hundred and seven articles (60%) with mention of IAEA underline that the organisation plays a major role representing the weight of the expertise the organisation carries. It is the communication "success" of the IAEA that it has managed to thematise the coverage of these topics. Similarly, in 64 out of 107 articles a major risk element was presented, implying losses of lives, major material damage and long-term negative impacts. An overwhelming majority of the articles with a major risk focus (47 out of 64, 70%) granted the IAEA a central role in the coverage. This implies acknowledging the competence and expertise of the organisation in critical risk situations. In most of the other articles, the IAEA has a role similar to other actors: representatives of the Ukrainian government, army or nuclear energy company, or a spokesperson for the Russian government. In most articles, the IAEA perspective is quoted to increase the weight of unbiased, technology and safety-focused content (Table 11.3).

The analysis identified the key risk communication frames as follows.

Table 11.3 Distribution of coverage of Ukraine articles, according to risk perception and role of IAEA

RISK PERCEPTION IN THE ARTICLE		*ROLE OF IAEA IN THE ARTICLE*		
		minor	*medium*	*central*
Minor	13	3	5	5
Medium	26	9	7	10
Major	64	6	11	47
Mixed	4	2	0	2
TOTAL	107	20	23	64

The analysis identified the key risk communication frames as follows.

Emphasis on complexity and seriousness of risks

IAEA communication regularly demonstrate that in a military conflict, the growing occurrence of risks includes nuclear-related risks, as well. While the hazards are not directly caused by nuclear power production activities, risks might impact them significantly, including the peaceful and safe operations of the plants. The complexity of nuclear power is evidenced by the range of risks that emerge especially under the critical conditions in Ukraine. Emphasising the grave nature of risks is common in the IAEA releases and news articles share the same concerned tone. In certain instance IAEA refers to the highest possible risks, to draw attention to the probability and impact of such a hazard, and imply the unique role IAEA has, through its competence, in attenuating this risk.

> *I remain deeply concerned about nuclear safety and security at the plant, both when it comes to its vulnerable off-site power supplies – which can be affected by attacks far away from the site – and the more direct military risks it is facing, potentially undermining the principles that I set out at the United Nations Security Council in May." – Update 197 – IAEA Director General Statement on Situation in Ukraine, 26 November 2023*

> *"The reports I receive from our experts indicate that the explosions occurred some distance away from the Zaporizhzhya Nuclear Power Plant. Nevertheless, I remain deeply concerned about the possible dangers facing the plant at this time of heightened military tension in the region. Whatever happens in a conflict zone wherever it may be, everybody would stand to lose from a nuclear accident, and I urge that all necessary precautions must be taken to avoid it happening," Director General Grossi said. – Update 182 – IAEA Director General Statement on Situation in Ukraine. – 8 September 2023*

Focus on technical competence in a politically sensitive situation

IAEA press releases reflect a matter-of-fact, strong operational focus. The interest of the organisation lies in supporting the safe operations of nuclear facilities. In a way the "business-as-usual" style of the press releases imply that the organisation knows what it speaks about and what is deals with. In the press releases, the organisation regularly underlines its technical capabilities, through which the operational safety of the nuclear power plants can be ensured. The choice of topics and the factual wording of the releases intend to calm the situation as well as underline the technical competence of the IAEA.

> *In a statement, Grossi said their presence was "indispensable to help reduce the risk of a nuclear accident. Our courageous experts – working closely with the plant's operating staff – are providing technical advice and monitoring the situation in extremely difficult and challenging circumstances," he added. – Reuters, 2 Marc, 2023*

> *Russia urged the International Atomic Energy Agency on Friday to ensure Ukraine does not shell the Zaporizhzhia nuclear power plant, saying it was otherwise operating safely. – Reuters, 23 June 2023*

Demonstration of confidence and control

In close support to the previous framing, the press releases regularly cover actions IAEA is currently taking or planning, which results in the demonstration of confidence in knowing what they are doing. This again contributes to the positioning of the situation as a technical issue. The press releases speak about the organisation and its representative in a confident manner, implying they know what they are speaking about and doing. The press releases focus on actions the IAEA is taking or is urging other to take. Difficult situations are presented together with actual or proposed decisions, again indicating that the organisation is in a leading position to solve complex nuclear matters.

> *The head of the U.N. atomic energy agency said on Thursday that ensuring water for cooling was a priority of his visit to the Russian-controlled Zaporizhzhia nuclear plant in Ukraine, adding that the station could operate safely for "some time". – Reuters, 15 June 2023*

Evidencing political neutrality

IAEA takes the internationally approved political status quo as the guiding line. Being a UN aligned organisation IAEA pursues acceptance of international law, including the respect of borders and national sovereignties. It considers the area under Russian occupation as part of Ukraine and even avoids referring to "Eastern Ukraine" as the location of the Zaporizhzhia Nuclear Power Plant (ZNPP). While its political neutrality is demonstrated in the wording of its press releases, its actions are questioned by the opposing countries with the aim to exercise pressure on its actions. Russia questions the IAEA's technical competence and raises doubts regarding its political neutrality.

> *"It is very, very important that we agree on the fundamental principle that a nuclear plant should not be attacked under any circumstances, and it also shouldn't be used to attack others... A nuclear accident with radiological consequences will spare no one" – Update 152 – IAEA Director General Statement on Situation in Ukraine, 30 November 2023*

> *"The IAEA has neither statutory, nor technological, nor other capabilities to prevent a nuclear catastrophe in the event of attacks on nuclear power plants," Renat Karchaa, an adviser to the CEO of Russian state energy organisation Rosenergoatom, told Russian state TASS news agency on Tuesday. "Therefore, from this point of view, the presence of IAEA inspectors at all nuclear power plants is ineffective and is in a greater part politically motivated." – Reuters, 17 January 2023*

Credible personal communication

Rafael Grossi, the IAEA Director General regularly appears as a key communicator in the IAEA press releases. The releases usually cover his travels, meetings, actions and messages in Ukraine and about Ukraine. Linking personal credibility with organisational credibility is a demonstration of his commitment to the cause of nuclear safety. His communication is closely tied to the organisational communication of the IAEA, demonstrated for example by the standard blue IAEA jacket he is wearing in every appearance. On several occasions, he praises his team of nuclear experts that do all the dangerous and responsible work on the fields.

> *"An IAEA team including the agency's Director General Rafael Grossi visited the plant on Thursday, crossing the front line to get there. A small team of IAEA experts based at the plant was also rotated out and replaced." – Reuters, 16 June 2023*

In overall terms, the IAEA releases reflect a clear drive to de-escalate the political, military and technological tensions. IAEA positions itself as a politically neutral actor whose only interest, and mandate, is in supporting the secure operations of the nuclear power plants in the field. Due to the nature of the conflict and most of the affected geographical area being a military zone, the IAEA often appears to be the only credible source of information on nuclear matters. The style and the content of press releases reflect confidence and support the relevance of the propositions IAEA has made to secure safe operations.

International

In 2023, IAEA issued 85 press releases about international (non-Ukraine-related) topics. These press releases covered 38 countries and 5 cooperation partners. The biggest number of press releases referred to Japan, with a focus on the release of treated water into the ocean at Fukushima. This was followed by three releases each about Belgium, the Czech Republic and Iran. In terms of their content, most pieces referred to official IAEA visits, revision of country nuclear regulations and information about the planned expansion of nuclear energy production capacities.

In the same period, the four news media published 149 articles on international (non-Ukraine related) topics with the mention of the IAEA. Reuters was the most active and Politico was the least active media actor in the field. Publication of news articles was shared between the media outlets as follows:

- Al-Jazeera: 13 articles
- Euronews: 12 articles
- Politico: 4 articles
- Reuters: 120 articles

In the case of international news, news media followed a robust filtering function, resulting in a significant difference from what IAEA thought was of interest to issue press releases about. News media articles covered 20 countries, compared to the 38 countries the IAEA communicated about. Almost 50% of the countries covered by the IAEA releases were not of interest to the media. On the other hand, news media referred to the IAEA in several articles about countries that had not been covered by IAEA press releases, including China, Israel, and North Korea. Of the IAEA press releases covering collaboration with international organisations, none were

Table 11.4 Focus countries and organisations in the international news media articles and IAEA press releases

	Number of articles – international	*Number of IAEA press releases – international*	*Article/press release*
TOTAL PIECES	**149**	**85**	**1.75**
Number of countries covered, of which:	20	38	
Iran	46	3	15.33
Japan	41	16	2.56
Republic of Korea	10	1	10
China	6	0	-
Libya	6	0	-
Articles covering other countries (with each country having less than four mentions):	40	65	
Article or release/country	7.45	2.23	
Number of organisations covered	0	5	

picked up by the news media. Not only did the media pursue a strong selection in terms of countries, but it focused dominantly on two topics. The discharge of the treated water at Fukushima turned into a topic of global relevance, and attracted significant coverage not only in direct relation to Japan but about reactions in China, North Korea, Korea, and the Philippines, as well. The other major topic with outstanding coverage was related to the international talks and diplomatic efforts about the Iranian uranium enrichment activities. Beyond these Asian countries, only one further country proved interesting to international media; a significant number of articles covered Libya referring to the ten drums of uranium ore that went missing. The article/press release ratio indicates how the media resonated with the IAEA press releases. In comparison, the cases from Japan and Iran generated a similar number of news articles (41 vs 46), however, the IAEA issued a significantly higher number of releases about Japan (16) than about Iran (3) (Table 11.4).

Risk communication frames internationally

Communication frame analysis follows the approach of the IAEA which develops and pursues national level development and support activities, reflecting the individual level of needs. Looking at frames in a country specific way consequently aligns with the national focused approach of the IAEA.

Japan

Focus on technical competence in a tense situation

The IAEA underlines that the preparation and the execution of letting treated seawater back to the sea at Fukushima over a couple of years is a fully controlled process. IAEA maintains its position despite the criticisms. News articles however regularly refer to public and political opposition to this plan over the whole East Asian region. Politicians, individuals and organisations claim that the decision was made without due consideration to all factors. Opponents directly or indirectly question the competence of the IAEA in agreeing to the plan of releasing contaminated water back into the sea. Media coverage indicates that the IAEA has not managed to convince the wider global public about the safety of this operation.

> *The International Atomic Energy Agency (IAEA), the U.N. nuclear watchdog, gave the plan* (the release of treated water into the sea) *a green light in July, saying it met international standards and the impact on people and the environment would be "negligible". – Reuters, 24 August 2023*

> *"The Japanese side should not cause secondary harm to the local people and even the people of the world out of its own selfish interests," Foreign Ministry, China, Reuters, 24 August 2023*

> *"The Fukushima nuclear disaster is not over. This time only around 1% of the water will be released. From now on, we will keep fighting for a long time to stop the long-term discharge of contaminated water." 71-year-old protester, Jun Iizuka, Reuters, 24 August 2023*

Expanding science and technology communication on behalf of the IAEA

IAEA strives to consolidate its position of supporting the release of contaminated sea water by bringing external experts on board. Broadening of the expert base from various areas, all supporting the decision about the treated water, serves the consolidation of the decision of the IAEA. The news media took notice of this change in communication, and in their articles, they matched experts' opinions with those of the public and the political forces in the region. Opening the communicator base by the IAEA was met with a similar broadening of the group of people commenting on the decision.

> *"There is an understandable perception that all radioactive materials are always and everywhere dangerous... but not all radioactive materials are dangerous," says Tony Irwin, an honorary associate professor at the Australian National University. – Euronews, 22 August 2023*

> *The IAEA also released a statement saying its independent on-site analysis had confirmed the tritium concentration was far below the limit. "There are not going to be any health effects... There is no scientific reason to ban imports of Japanese food whatsoever," said Geraldine Thomas, former professor of molecular pathology at London's Imperial College., Reuters, 24 August 2023*

Iran

Emphasis on the shared goal of cooperation

The central point of IAEA involvement has been to oversee the Iranian nuclear capabilities and uranium enrichment activities. The IAEA intends to put those points in the focus where the various actors agree. The IAEA is firm but cautious in its use of words as it strives to maintain the process of negotiations. Clear wording is part of its practice to demonstrate they know what they speak about, and consequently contribute to their manifestation of competence. Simplification of the wording of IAEA induces similar simple messaging from the other actors, for example, Iran and the USA. Consequently, simplified communication leads to a stronger appearance of critical issues and conflict.

> *"Iran's stance is not only unprecedented, but unambiguously contrary to the cooperation that is required." – Rafael Grossi, IAEA DG, Al-Jazeera, 15 November 2023*

> *"Iran's nuclear escalation is all the more concerning at a time when Iran-backed proxies continue their dangerous and destabilising activities in the region". – White House National Security Council spokesperson, Al-Jazeera, 27 December 2023*

> *"We did nothing new and are doing the same activities according to the rules." – Mohammad Eslami, Iran's Atomic Energy Chief, Al-Jazeera, 27 December 2023*

Underline technical competence and sense of achievement

The IAEA positions itself as an organisation with outstanding competence in the field of nuclear technology. Competence is underlined through the presentation of the specific steps the organisation has taken to ensure the operational safety of nuclear activities. The use of words commonly refer to professional and scientific activities the IAEA has been doing or is supporting.

> *"We have agreed that those [cameras and other equipment] will be operating again. These are not words. This is very concrete." – Rafael Grossi, IAEA DG, Al-Jazeera, 4 March, 2023*

Use personal communication to increase credibility

Personal and organisational communication align closely as IAEA Director General, Rafael Grossi frequently communicates in person. This includes the use of his personal social media channels, most importantly, X (formerly Twitter). The IAEA press releases are often connected to Grossi's personal communication channels. Through aligning organisational and individual communication, the IAEA aspires to gain further credibility to its communication and actions.

> *"The IAEA is aware of recent media reports relating to uranium enrichment levels in Iran," the agency wrote on Twitter early on Monday. "Director General @rafaelmgrossi is discussing with Iran the results of recent Agency verification activities and will inform the IAEA Board of Governors as appropriate." – Al-Jazeera, 20 February, 2023*

In summary, the IAEA international releases are focusing on developments related to nuclear issues in individual countries. While the IAEA intends to keep the national focus of its activities, in reality this interpretation is not always met by the public. Media news, for example, provides wide spread coverage of the regional reactions to the release of contaminated seawater at Fukushima. The IAEA intends to limit risk communication to Japan and present the case as a technical issue. Media, however, alters the risk perception by looking at it from a wider perspective in geographical, social, economic and political terms. In the case of Iran the IAEA issues only a limited number of press releases to cover its role in overseeing the Iranian uranium enrichment activities. It positions the case as an example for international cooperation, however, media does not seem to agree with this framing. The global interest in the issue is significant and focuses on the conflict element, evidenced by a large number of news media articles on the topic, many of them with a significant risk element. Most of the articles refer to the case of Iran as an example of political tensions.

Discussion

Textual analysis of the IAEA press releases and the news media articles on Ukraine demonstrate a dual picture. On the one hand, press releases highlight substantial risk factors, for example, the escalation of military activities in the imminent vicinity of a nuclear power plant or the risk of not providing a sufficient amount of water for reactor cooling. On the other hand, the IAEA is quick to attenuate risk factors by putting in the focus its presence in the field, personal engagement and technical expertise. IAEA press releases focus on what the organisation, and personally its Director General, Rafael Grossi does concerning imminent risks. In several releases, highlighting the risk together with the action, the IAEA creates a feeling of control over the situation.

In the case of other countries, the analysis of the IAEA press releases and the role of news media show a more diverse picture. Media fulfils its filtering and altering function by defining and representing different priorities than IAEA. First, the media simply ignores a significant part of the IAEA communication and neglects the press releases that are about formal issues, have no news value or refer to areas that present no interest to their perceived audience. Instead, news media focuses on topics that have a strong embedded risk content. This explains their clear choice of focusing on the developments in the aftermath of the Fukushima disaster in Japan, affecting and involving all countries in East Asia. In Iran, the organisation represents itself as the key actor controlling the uranium enrichment activities, however, the IAEA perspective is often complemented or even opposed by the news articles. These articles present the IAEA being caught in a political conflict between various countries and actors. The case of the missing uranium ore drums in Libya is a similar case, where despite the IAEA not covering the topic at all, news media found it of interest. These three countries and topics dominate the international news media coverage.

The comparison of the coverage of Ukraine and the other countries presents a significant difference. In the case of Ukraine, the IAEA's role as the "global nuclear watchdog of the UN" is well respected by the media. News articles most commonly reflect the IAEA viewpoint and very often underpin their findings with information from the IAEA. The international coverage shows a more diverse picture where much of the information from the IAEA is complemented or even overruled by information from other sources. Risk perception seems to be the main driver for the media. The higher and more complex they see the risks, the more likely it is they are referring to non-IAEA sources in their articles.

Based on the "filter and alter" function, the media acknowledges the role and the technical competence of the IAEA. The organisation is regularly noted as the "United Nations' nuclear watchdog", and Rafael Grossi is quoted and often pictured in the majority of articles. While, especially on political grounds, there are occasional doubts raised about its role, these doubts are always balanced with supportive claims from other sources.

Especially in the case of Fukushima, however, the news media amplifies the risks significantly. By regularly representing views that challenge and even confront the approved process of seawater treatment, news media takes an active role. The emergence of opposing views and the high level of criticism is related to Japanese authorities carrying a "stigma" for their perceived responsibility in mishandling the Fukushima case in 2011. Social Amplification of Risk represents in this case the complex and contradictory views about this process. The reaction and protest from around the region can also be explained by the "ripple" effect, where one development leads to the next one and ultimately contributes to the strengthening of the controversy.

News media coverage attests to the social amplification station function, by demonstrating first the selection of the topics the media makes in the filtering stage, followed by the contextualisation in a format appropriate to the given media in the altering stage. In the first stage, the media *filters* information by deselecting topics that they consider of non-interest. The textual analysis evidences that news media has little or no interest in topics that refer to IAEA internal affairs and concern meetings the content of which are unknown or are overtly technical. IAEA press releases on these topics are simply ignored by the media. In the next stage, the contextualisation of the press release information is taking place. This includes *altering* the original content, and adapting it to the specific needs, focusing on those parts of the information that are relevant to the media. News media make this choice by assigning the level of importance and relevance of the topics according to the perceived audience of the media. Very often this includes referring to a broader range of information sources, some of which are in contradiction with what the IAEA press release claimed.

In the case of media, the selection of topics and their interpretation form the two stages of filtering and altering. In the Social Amplification of Risk Framework, these two stages are identified as the role of media of social amplification and attenuation of information. The current study confirms the ongoing relevance of the Social Amplification of Risk Framework, and especially, the interpretation of media channels as "social amplification stations" in the risk awareness processes. While the research confirms that the authority and credibility of the IAEA is acknowledged by the media, news media channels maintain their role and integrity by maintaining their unique status in the Social Amplification of Risk Framework. This leads to confronting, overwriting or even neglecting the views of the IAEA in certain cases. Fulfilling the social amplification station role is consequently a dynamic practice, where decisions about amplification or attenuation of risks is done on a case by case basis.

The notion of the role of each recipient (Kasperson et al., 1988) is confirmed by each news media following its own social amplification practice. This study confirms the limitations of risk communication focusing on technical aspects only (Arvai & Rivers, 2013), as seen primarily in the limited effect of IAEA press releases on Fukushima achieved. It is confirmed that the technical aspect of risk dominates the discourse and the social, political and cultural dynamics that influence risk management and risk perception are underappreciated. In their view, risk communication is ineffective if it focuses on providing additional technical information only. The presence of "stigma" towards the Japanese government, authorities and TEPCO, as described by de Groot et al., 2013, and Kusumi et al., 2017 are also confirmed in this study. Media criticality towards the IAEA viewpoints reflecting the views of the Japanese government is induced by stigma from handling of the Fukushima disaster in 2011.

Conclusions

In 2023, the IAEA demonstrated the status of "approved supplier" of credible and competent information. This status is especially apparent in its communication about Ukraine. The analysis of press releases and news articles confirms that the IAEA and its personnel, headed by Director General Rafael Grossi manifest commitment to the cause on the ground and provide regular and accurate information. Communication focuses on technicalities and puts ensuring nuclear safety in the focus.

This credibility is questioned by the media in the case of the international releases. In these cases, information from the IAEA is considered one of the important sources of nuclear information but is extended significantly with other viewpoints and sources. The news articles are created through a significant level of filtering and altering of the IAEA information. The news media, evidencing its social amplification station role, covers the developments of these processes.

Through the involvement of a wide source of information, its risk messages are amplified. In the case of the articles on Japan, their content furthermore reflects a loss of confidence in the Japanese authorities and a reduced level of confidence in the competency of the information provided by the IAEA.

In response to the original research question, the analysis validates that filtering and altering of information form two separate steps of news media activities. This is evidenced by the way the media handled the press releases of the IAEA; first, they filtered and then altered the information of the IAEA press releases. It can be furthermore confirmed that both amplification and attenuation of risks might happen in any of the two stages. In the case of Ukraine, the IAEA served as a major source of credible information and the news media attenuated the risk content of the messages. The case of the Fukushima plant however leads to a different direction; when the IAEA was not the only source of information and its credibility was challenged by other players, the news articles contributed to an amplification of risks. This case confirms that especially when the original source downplays the risk content of the message, news media "stations" might significantly counterbalance this by amplifying it. The robust and dynamic presence of attenuation and amplification of risk messages confirms the relevance and the validity of the Social Amplification of Risk Framework, and the role of media as social amplification stations, within.

Limitations of the study

The research had to face certain limitations. The sample sizes did not allow for detailed quantitative analysis, numerical findings therefore serve as indication only. In selecting the media for study, access to information had to be considered and this may have limited the range of sources the study could use. Finally, the last stage of the process, a detailed study of actual public risk perceptions could not be part of the research, due to the planned volume of the chapter. In its current form, however, it can provide a strong basis for further investigation.

Acknowledgements

The author would like to thank Dr. Dorottya Egres, Budapest University of Technology and Economics, for their contribution to the formulation of the key concept and the adaptation of the SARF framework, and Judit Sarlos and David Sarlos for their committed work in coding the texts and identifying key text corpora for study.

References

Angelique, H., & Cunningham, K. (2006). Media framing of dissent: the case of initial anti-nuclear protests following the Three Mile Island accident. *The Australian Community Psychologist, 18*, 42–57.

Arvai, J., & Rivers III, L. (Eds.). (2013). *Effective risk communication* (1st ed.). Routledge. https://doi.org/10.4324/9780203109861

Assmann, J. (2008). Communicative and cultural memory. In A. Erll, & A. Nünning (Eds.), *Cultural memory studies: an international and interdisciplinary handbook* (pp. 109–118). Berlin/New York: de Gruyter.

Bakir, V. (2010). Media and risk: old and new research directions. *Journal of Risk Research*, *13*(1), 5–18. https://doi.org/10.1080/13669870903135953

Beck, U. (1999). *World risk society*. Cambridge, Polity Press.

Culley, M.R., Ogley-Oliver, E., Carton, A.D. & Street, J.C. (2010), Media framing of proposed nuclear reactors: an analysis of print media. *Journal of Community and Applied Social Psychology*, *20*, 497–512. https://doi.org/10.1002/casp.1056

Doyle, J. (2011). Acclimatizing nuclear? Climate change, nuclear power and the reframing of risk in the UK news media. *International Communication Gazette*, *73*(1–2), 107–125.

Egres, D., & Sarlos, G. (2023). Nuclear perceptions from radioactive blue to sustainable green: the EU taxonomy as reflection of a divided public. *Journal of Public Affairs*, *24*(1), e2901. https://doi.org/10.1002/pa.2901

Entman, R.M. (1993). Framing: toward clarification of a fractured paradigm. *Journal of Communication, 43*(4), 51–8.

Fischhoff, B., Lichtenstein, S., Slovic, P., Derby, S.L. & Keeney, R.L. (1981). *Acceptable risk*. Cambridge: Cambridge University Press.

Fischhoff, B., Slovic, P., Lichtenstein, S. Read, S. & Combs, B. (1978) How safe is safe enough? A psychometric study of attitudes towards technological risks and benefits. *Policy Sciences*, *9*, 127–152. https://doi.org/10.1007/BF00143739

Flynn, J. (2003). Nuclear stigma. In: *The social amplification of risk* (pp. 326–352). Cambridge: Cambridge University Press.

Gamson, W. & Modigliani, A. (1989). Media discourse and public opinion on nuclear power: a constructionist approach. *American Journal of Sociology*, *95*, 1–37.

Goatcher, J., & Brunsden, V. (2011). Chernobyl and the sublime tourist. *Tourist Studies*, *11*(2), 115–137. https://doi.org/10.1177/1468797611424956

de Groot, J.I.M., Steg, L. & Poortinga, W. (2013). Values, perceived risks and benefits, and acceptability of nuclear energy. *Risk Analysis*, *33*, 307–317. https://doi.org/10.1111/j.1539-6924.2012.01845.x

Hagen, B., Opejin, A. & Pijawka, K.D. (2022). Risk perceptions and amplification effects over time: evaluating fukushima longitudinal surveys. *Sustainability*, *14*(13), 7896. https://doi.org/10.3390/su14137896

Hara, T. (2013). Social shaping of nuclear safety: before and after the disaster. In R. Hindmarsh (Ed.), *Nuclear disaster at Fukushima Daiichi: social, political and environmental issues* (pp. 22–40). New York, NY: Routledge.

IAEA (2024a, 2024 January 8). *Country Nuclear Power Profiles*. https://cnpp.iaea.org/public/

IAEA (2024b, January 6). *The IAEA Mission Statement*. https://www.iaea.org/about/mission

Kasperson, R.E., Renn, O., Slovic, P., Brown, H.S., Emel, J., Goble, R., Kasperson, J.X. & Ratick, S. (1988). The social amplification of risk: a conceptual framework. *Risk Analysis*, *8*, 177–187. https://doi.org/10.1111/j.1539-6924.1988.tb01168.x

Kasperson, R.E., Webler, T., Ram, B., & Sutton, J. (2022). The social amplification of risk framework: new perspectives. *Risk Analysis*, *42*, 1367–1380. https://doi.org/10.1111/risa.13926

Katz-Rosene, R.M. (2021). 'Don't think of Fukushima!': the ethics of risk reframing in 'Nuclear for Climate'. *Communications, Ethics, Policy & Environment*, *24*(2), 164–186, DOI: 10.1080/21550085.2020.1864800

Kinsella, W.J. (2005). One hundred years of nuclear discourse: four master themes and their implications for environmental communication. *Environmental Communication Yearbook, 2*, 49–72. London: Routledge.

Kristiansen, S. (2017a). *Media and risk: a phase model elucidating media attention to nuclear energy risk* (S. Kristiansen; Vol. 5) [Universitätsverlag Ilmenau]. https://nbn-resolving.org/urn:nbn:de:gbv:ilm1-2017100025

Kristiansen, S. (2017b). Characteristics of the mass media's coverage of nuclear energy and its risk: a literature review. *Sociology Compass, 11*, e12490. https://doi.org/10.1111/soc4.12490

Kristiansen, S., Bonfadelli, H., & Kovic, M. (2016). Risk perception of nuclear energy after Fukushima: stability and change in public opinion in Switzerland. *International Journal of Public Opinion Research, 30*(1), 24–50. https://doi.org/10.1093/ijpor/edw021

Kusumi, T., Hirayama, R., & Kashima, Y. (2017). Risk perception and risk talk: the case of the Fukushima Daiichi nuclear radiation risk. *Risk Analysis*, *37*(12), 2305–2320. https://doi.org/10.1111/risa.12784

McCombs, M.E., & Shaw, D.L. (1972). The agenda-setting function of mass media. *The Public Opinion Quarterly*, *36*(2), 176–187. http://www.jstor.org/stable/2747787

Miller, D.S. (2016). Public trust in the aftermath of natural and na-technological disasters: Hurricane Katrina and the Fukushima Daiichi nuclear incident, *International Journal of Sociology and Social Policy*, *36*(5/6), 410–431. https://doi-org.roe.idm.oclc.org/10.1108/IJSSP-02-2015-0030

Novikau, A. (2017). What is "Chernobyl Syndrome?" The use of radiophobia in nuclear communications. *Environmental Communication*, *11*(6), 800–809. https://doi.org/10.1080/17524032.2016.1269823

Pidgeon, N.F. & Barnett, J. (2013). *Chalara and the social amplification of risk.* London: DEFRA. https://www.gov.uk/government/uploads/system/uploads/attachment_data/file/200394/pb13909–chalara-social-amplification-risk.pdf

Pidgeon, N.F., Lorenzoni, I. & Poortinga, W. (2008). Climate change or nuclear power—No thanks! A quantitative study of public perceptions and risk framing in Britain. *Global Environmental Change, 18*(1), 69–85. https://doi.org/10.1016/j.gloenvcha.2007.09.005

Van Der Pligt, J. (1985). Public attitudes to nuclear energy: salience and anxiety. *Journal of Environmental Psychology*, *5*, 87–97. https://doi.org/10.1016/S0272-4944(85)80040-2

Rantasila, A., Sirola, A., Kekkonen, A., Valaskivi, K. & Kunelius, R. (2018). #fukushima Five years on: a multimethod analysis of twitter on the anniversary of the nuclear disaster. *International Journal of Communication*, *12*, 928–949, http://ijoc.org/index.php/ijoc/article/view/6475/2277

Renn, O. (2010). The contribution of different types of knowledge towards understanding, sharing and communication risk concepts. *Catalan Journal of Communication & Cultural Studies*, *2*(2), 177–195.

Rieu, A.M. (2013). Thinking after Fukushima. Epistemic shift in social sciences. *Asia Europe Journal*, *11*(1), 65–78.

Sarlos, G. (2014). *Risk and benefit perceptions in the discourse on nuclear energy*. LAP Lamberts.

Sarlos, G., & Fekete, M. (2019). Seeing the unseen: the Chernobyl imprint on the Hungarian public sphere. *Journal of Public Affairs*, *19*(4). https://doi.org/10.1002/pa.1931

Slovic, P., Flynn, J., Mertz, C.K., Poumadère, M., & Mays, C. (2000). Nuclear power and the public. In O. Renn & B. Rohrmann (Eds.), *Cross-cultural risk perception. Technology, risk, and society* (vol. 13). Springer. https://doi.org/10.1007/978-1-4757-4891-8_2

Vainio, A., Paloniemi, R., & Varho, V. (2017). Weighing the risks of nuclear energy and climate change: trust in different information sources, perceived risks, and willingness to pay for alternatives to nuclear power. *Risk Analysis*, *37*(3), 557–569. https://doi.org/10.1111/risa.12640

Visschers, V.H.M., & Siegrist, M. (2013). How a nuclear power plant accident influences acceptance of nuclear power: results of a longitudinal study before and after the Fukushima disaster. *Risk Analysis: An Official Publication of the Society for Risk Analysis*, *33*(2), 333–47. https://doi.org/10.1111/j.1539-6924.2012.01861.x

Vossen, M. (2020). Nuclear energy in the context of climate change: a frame analysis of the Dutch print media. *Journalism Studies*, *21*(10), 1439–1458. https://doi.org/10.1080/1461670X.2020.1760730

Weart, S.R. (2012). The rise of nuclear fear. Harvard University Press. https://doi.org/10.4159/harvard.9780674065062

World Nuclear Association (2023, November). *Nuclear Power in the World Today*. https://world-nuclear.org/information-library/current-and-future-generation/nuclear-power-in-the-world-today.aspx#:~:text=The%20first%20commercial%20nuclear%20power,the%20world's%20low%2Dcarbon%20electricity.

Zhukova, E. (2016). From ontological security to cultural trauma: the case of Chernobyl in Belarus and Ukraine. *Acta Sociologica*, *59*(4), 332–346. https://doi.org/10.1177/0001699316658697

12 Journalist–source relations in Russian environmental journalism

Olga Dovbysh and Mika Perkiömäki

Introduction

Sourcing and the relationship between journalists and news sources are important in highly contested issues such as climate change or environmental crises. Previous research examining the nature, influences, and difficulties involved in reporting on environmental issues has underscored an increasing number of news sources all vying to establish their perspective. The trend towards politicisation of the topic started in the late 1980s when politicians began to increasingly influence the agenda and now dominate sources globally (Anderson, 2017). Recent research shows that journalists in both democracies and autocracies tend to refer to mainstream political/government viewpoints and positions, and are heavily reliant on official sources in environmental and climate reporting (Comfort et al., 2020; Ejaz et al., 2022). Politicisation is not the only factor shaping the relationship between journalists and sources worldwide. Others include the impact of digital technologies, the growing significance of public relations practitioners, and the rise of citizen journalism (Anderson, 2017).

Despite the evidence of similar trends in the relationship between journalists and news sources across the world, national context and political regime still matter (Comfort et al., 2020). Social systems and cultural contexts exert an influence on news construction processes (Shoemaker & Reese, 2014), which in practice means that rather than reflecting a shared global problem, environmental and climate journalism reflects national journalistic cultures (Brossard et al., 2004; Vu et al., 2019). There is a long tradition of depicting differences in national climate journalism between the highly developed, fossil fuel-dependent Global North, and the less developed, more vulnerable to the effects of climate change Global South (Schäfer & Painter, 2021). However, this approach implies significant overgeneralisation, and researchers are calling for more nuanced studies on climate change journalism that take into account the country's political regime and media system, as well as the country's relative responsibility for, and vulnerability to climate change, for instance (Comfort et al., 2020).

Russia occupies a specific position in the dichotomy between the Global North and the Global South. Russia is often understood in relation to the West (Sakwa, 2021), especially in the context of the ongoing Russo-Ukrainian war. The Russian

DOI: 10.4324/9781003479550-15

economy and political system are highly dependent on the extraction and export of oil and gas, and Russia has been described as a "hydrocarbon superpower" (Tynkkynen, 2019; Bouzarovski & Bassin, 2011). Hence, climate change has been repeatedly portrayed by the Russian leadership as beneficial for Russia because it would enable agriculture to expand into new areas and lead to the melting of the Northeast Passage. Moreover, Russian society is quite indifferent to climate and environmental issues: only 1% of respondents mentioned climate and natural disasters as worrying issues, while garbage and landfill problems and water pollution were among the most pressing ecological concerns, cited by more than 50% of respondents (FOM, 2024). Given the increasing authoritarianism and severe restrictions on media freedom, environmental journalism in Russia serves as a crucial example of how the environmental agenda is constructed and narrated by the media in autocracies.

In this chapter, we focus on sourcing as a core element of the journalistic work affecting the construction of the environmental agenda. Using the Hierarchy of Influences model (Shoemaker & Reese, 2014) and data from in-depth interviews with Russian environmental journalists, we explore how journalist-source relations in environmental and climate reporting in Russian media are organised under the influence of various forces and factors. To this end, we aim to address the more general issue of the (re)production of environmental discourses in the media in an authoritarian regime and the agency of media professionals in this process.

The rest of the chapter is structured as follows. First, we discuss the epistemology of sources in journalism and environmental journalism in particular. Next, we operationalise sourcing as a factor of influence and elaborate on the Hierarchy of Influences model to analyse this influence at different levels. We then present our data and methodology. In the results section, we explore journalist-source relations in climate and environmental reporting in Russia. The chapter concludes with a discussion of our findings on the influence of sourcing on the environmental agenda in authoritarian regimes like Russia.

Sources and data in environmental journalism

Examining the influence of sources on journalistic decisions reveals broader issues of social dominance and legitimacy (Carlson, 2009). Objectivity, a guiding paradigm of Western journalism, is inextricably linked to sources. Epistemologically, sources serve as an essential form of evidence (Tuchman, 1978). Journalists use their own professional authority to rework information attributed to sources into facts that require no further verification. At the same time, journalistic reliance on sources goes hand in hand with the question: Whose version of the world is being represented?

Although journalistic norms emphasise professional autonomy and the inclusion of a plurality of alternative voices, elite sources are notably dominant and routinely privileged by media outlets due to their authoritative position and substantial resources for gaining media attention (Anderson, 2017). As environmental reporting, and the climate change debate in particular, has become increasingly politicised in recent decades, government sources have significantly increased their activities and resources dedicated to attracting media attention. In environmental reporting, as in other beats, government sources now tend to have the most say in both the Global North (Schäfer & Painter, 2021) and the Global South (Comfort et al., 2020; Huan, 2024). Environmental NGOs (ENGOs) and other activist groups are less likely to be seen as neutral and reliable sources, but are more prominent in crisis situations when official sources may be slower or adopt a reactive position (Anderson, 2017). In an authoritarian context, given the limited media pluralism, the dominance of government sources further strengthens the existing power hierarchy and undermines alternative voices, thus limiting the number of contested truths in the media (Atton & Wickenden, 2005). This also affects the way in which facts related to environmental and climate issues are constructed in the media: the case of China demonstrates that the news media use climate change as an issue to enhance the political performance of the leadership (Guo et al., 2023).

Sourcing as an influence in environmental journalism

In this chapter, we use the Hierarchy of Influences (HOI) model (Shoemaker & Reese, 2014) to understand the dynamics of the interrelations between environmental journalists and sources in the Russian media. Prior studies have used the HOI model extensively to explore and assess the way in which different influences impact climate-related journalistic content (Ejaz et al., 2022; Comfort et al., 2020; Anderson, 2017; Takahashi et al., 2017; Duan & Takahashi, 2017). We employ the model to understand how sourcing in environmental journalism in contemporary Russia is shaped by different levels of influence.

The Hierarchy of Influences model is a theoretical framework for studying how news media content is produced and influenced. The model "takes into account the multiple forces that simultaneously impinge on media and suggests how influence at one level may interact with that at another" (Shoemaker & Reese, 2014, p. 1). These forces range from the micro to the macro level: *individual* characteristics of specific newsworkers, their work *routines*, *organisational*-level concerns, *institutional* issues, and larger *social systems*. The *individual* level of HOI considers the personal traits of journalists, the news values they espouse, the professional roles they take on, and other demographic features (such as gender, race, class) (Reese & Shoemaker, 2016). This level addresses the journalistic autonomy in making decisions, although they still operate within a web of constraints. Therefore, how individual journalists are shaped by, contribute to, and identify with their surrounding organisations are also part of this level (Reese & Shoemaker, 2016). The *routines* level refers to those patterns of behaviour that form the immediate structures of journalistic work. In HOI, routines are understood as reactions to the power exercised within organisations by establishing a pattern of practices that "serve the needs of the organization, adapt to requirements of information sources, control the workflow, and give it a meaningful structure" (Reese & Shoemaker, 2016, p. 399). The *organisational* level stems from the idea of news as an organisational product that had to be socially constructed. Hence, this level reflects how different parts of a media organisation, including the organisation's ownership, policies, goals, actions, rules, membership, interactions with other organisations, bureaucratic structure, economic viability, and its stability, shape the media outlet's journalistic product (Shoemaker & Reese, 2014). The *social institution* level considers how the various organisations involved in media work cohere into a larger institution. At the same time, the media institution is influenced by the way in which it enters into structured mutually dependent relationships with other major systemic players, including the state, public relations, and advertising (Reese & Shoemaker, 2016). The social system is the structure of relationships between people and the institutions they create. The *social system* level of influence includes influences from the system as a whole, including ideological, cultural, political, and economic forces, in the sense that they concern the meaning of the news in the service of social interests and power (Shoemaker & Reese, 2014).

The model proved useful when it came to unpacking the core concepts within communication research across this levels-of-analysis perspective (Reese & Shoemaker, 2016). By making the important distinction between structure and agency in media work, the model allows for an analysis of the operation of combined factors that are sorted along micro, meso, and macro levels. As such, in using the HOI model, we seek to address all layers influencing the journalist-source relationship in order to understand its complexity and importance for journalistic work and agenda-setting in general. Previous research has demonstrated different approaches to understanding sourcing in journalism and the factors that inform its complex and multi-layered nature. For instance, some address the journalists' own agency, practised through thinking, acting, and establishing knowledge when engaging with conflicting sources (Reich & Barnoy, 2021). Others focus on the changing roles of the sources in journalistic content (Laursen & Trapp, 2021). Finally, some research makes sense of sourcing as an act of domination by cultural power holders (Carlson, 2009). Even though some research on environmental journalism regards sourcing as a part of the routines level (Duan & Takahashi, 2017), we contend that it plays a more influential role, not only in influencing media coverage, but also in shaping a country's environmental journalism. Considering sourcing as a core element that shapes journalism and the mediated reality it produces, we argue that factors influencing sourcing as a professional practice, rooted not only in the media system but also in the country's political and social systems, make HOI an extremely relevant tool for this research.

Environmental and climate reporting in Russia

This chapter deals with environmental journalism, which can cover a wide range of environmental topics, such as waste management, pollution, and wildlife conservation. Climate change is a common theme in environmental journalism, and we consider climate journalism to be a subcategory of environmental journalism. Different environmental issues often overlap, and whether a story falls under climate journalism or another subcategory of environmental journalism depends on how it is framed. In the Russian context, for instance, extensive forest fires in Siberia can be framed as climate journalism because global warming increases their incidence rate, and the loss of forest increases the rate of global warming. However, they can also be treated as an environmental issue of their own because they are often caused by human activities other than climate change. Climate change is one lens through which the history and state of environmental journalism in Russia can be understood (Dovbysh & Perkiömäki, 2024). In this chapter, we use the notion of environmental journalism to cover all types of reporting on ecological, environmental and climate issues. Where necessary, we specifically highlight climate or other types of journalism as subcategories of a generalised notion of environmental journalism in Russia.

The authoritarian conditions in the media, politics, economy, and civil society contributed to keeping climate change on the sidelines of the social debate in Russia (Poberezhskaya, 2015; Yagodin, 2017; Beuerle, 2023). In the 1990s, the level of Russian media attention to climate change was very low, with the few reports usually focusing on its negative effects (Poberezhskaya, 2018). In the 2000s, the role of science in Russian climate reporting was contradictory (Wilson Rowe, 2009), with a rise in climate scepticism in the Russian media occurring between 2009 and 2012 (Poberezhskaya, 2018). In 2012–2013, climate denialism, conspiracy theories, logical inconsistencies and highlighting the positive effects of climate change for Russia were used to support the idea of Russia's national identity as a superpower based on fossil fuel self-sufficiency (Tynkkynen & Tynkkynen, 2018).

In late 2019, however, the political discourse began to reflect anxiety about a serious threat to Russia's future stemming from a warming climate (Gustafson, 2021, pp. 17–24). Echoing the earlier pattern, where changes in Russian state climate policy have influenced both the nature and the extent of media coverage about climate change (Wilson Rowe, 2009; Poberezhskaya, 2015; Ashe & Poberezhskaya, 2022), more concerned stories about climate change appeared with increasing frequency in the Russian media. This was accompanied by an increase in environmental reporting in general, with many newsrooms hiring environmental journalists or organising separate departments to cover the "green" agenda (Davydova, 2020). Several niche media specialising in environmental issues were launched. The late 2010s and early 2020s were also marked by increasing activity by grassroots environmental movements and international ENGOs, as well as ESG (environmental, social, and corporate governance) reporting among Russian businesses.

Despite the authoritarianism of the Russian regime and the unfree nature of the Russian media model, climate reporting has been relatively free from censorship (Ashe & Poberezhskaya, 2022). According to Davydova (2020), the same is true of environmental reporting in general, as it has been considered marginal and unimportant. The idea of environmental issues as "a little corner of freedom" in activist organisations, scientific institutions, literature, and the press existed as early as the Soviet period (Weiner, 1999).

The Russian media environment changed dramatically following the full-scale invasion of Ukraine in February 2022. The Russian authorities promptly introduced de facto war censorship, which led to over a thousand journalists leaving the country in order to escape the immediate danger of incarceration, and to continue their work censorship-free. Media outlets remaining in the country have to toe the government line, especially on the most sensitive issues. These changes have inevitably affected environmental and climate reporting. First, both the domestic and exile media are focusing more on the social consequences and political causes of the war, while environmental issues are becoming less visible on the agenda (Dovbysh & Perkiömäki, 2024). Second, environmental issues are becoming more politicised when it comes to Russia's political agenda and the work of international and domestic environmental NGOs. Third, the massive closure of open public data after February 2022 also affected access to environmental data. As many as 44 government agencies closed almost 500 datasets, largely related to military activities and their consequences (Esli byt' tochnym, 2023). However, some ecological datasets were also hidden, such as data on the volume of wastewater discharged into Lake Baikal and the Volga, information on the number and area of forest fires, upper-air data, and other datasets used by journalists in their work and investigations. All these changes would inevitably alter the way that Russian environmental journalists engage with their sources and how journalist-source relationships influence media content related to the topics. In the next section, we analyse these changes based on empirical data from in-depth interviews with journalists.

Data and method

This research is based on data from 18 semi-structured, in-depth interviews with journalists engaging with the environmental and climate agenda in Russian media outlets. When we conducted the interviews (mainly in the spring of 2022), most of the respondents were in Russia and only a few had recently relocated abroad, so all respondents reflected on their experiences of working in newsrooms in Russia. Respondents were enlisted using a snowball sampling technique to identify the type of media orientation (pro-state, oppositional), ownership structure (state-owned and commercial), and niches in environmental journalism (Davydova, 2020). The represented media include four business media, four opposition media, three state-owned news agencies and newspapers, two niche online media, two freelancers for various media, one commercial news agency, one regional newspaper and NGO, and one pro-state newspaper. The interviewees' areas of expertise were: ecology and business (5), environmental problems and society (4), climate change (3), various environmental topics (2), sustainable development (2), regional environmental issues (1), and environmental protests (1). Due to restrictions on travelling to Russia, the interviews were conducted via video calls in Zoom or Telegram. Interviews lasted between 49 and 165 minutes, with an average of 81 minutes and a median of 70 minutes. To guarantee anonymity, we are not publishing the names of the media for which the interviewees worked. For the same reason, the transcripts are not publicly available.

In the analysis, we employed a multistep approach (Kvale, 1996) to identify the key themes related to journalist-source relations and to understand how sourcing in journalistic work is subject to multiple levels of influence. First, we carefully read each interview, noting important relevant topics. We then identified condensed units of meaning and examined the connections and patterns among these units across the interviews, grouping them into the dominant themes in the respondents' narratives. Finally, the main themes were linked together according to HOI levels.

Russian environmental journalism and access to data and sources

Russian environmental journalists' reflections on their engagement with sources are consistent with their counterparts from other countries in their selection of sources (Comfort et al., 2020; Ejaz et al., 2022). Government sources and scientists are prioritised, followed by ENGOs and corporations. At the same time, the peculiarities of the national media model and journalistic culture influence their relationships with sources at all levels of the HOI model.

Individual level: A journalist's professional background and their involvement in the topic determine their engagement with sources at the individual level. Journalists with longer experience in environmental reporting claim a higher level of expertise in the topic and consequently exercise greater gatekeeping in the selection of sources. For issues such as climate change, where the selection of prominent experts is limited, seasoned journalists typically know all the key experts personally, as well as their opinions. As a result, the journalist independently decides who to give a voice to. Given that there are rather many climate sceptics among Russian scientists (Ashe & Poberezhskaya, 2022), journalists act as gatekeepers in managing the sceptical voices in the media.

Experience in the topic is an important factor in a journalist's ability to critically assess the credibility of sources. One of the respondents expressed their distrust of a scientific source, stating:

> Unfortunately, we're in an environment now where I can sit in front of a professor who should know everything, and as I'm just a journalist, I should basically believe him. But I don't trust him from the start because I know that this institute is often used to approve decisions made from above. That's why there are still problems with the academic environment.
>
> (R05, online business media)

Journalists' personal attitudes also influence their selection of sources. For example, one respondent (R18, online niche media) observed that in today's Russia, where the agency of environmental organisations has been significantly constrained, journalism is taking on a more activist role. Environmental journalists, they suggested, are expected to take on roles traditionally filled by ENGOs to some extent. This position clashes with that of another group of journalists who approach environmental issues from a more utilitarian perspective, both personally and professionally. For instance, a journalist working in the business media, primarily covering oil and gas corporations, adopted a more economy-driven perspective on climate change mitigation, without taking any activist stance:

> Up to now, all this [the developed countries' pursuit of decarbonisation] seems like a declaration aimed at developing certain technologies to make them economically attractive. If this were truly a global goal… to directly improve the climate... then many decisions, even in Glasgow [at the 2021 United Nations Climate Change Conference], would have been different. So far, it looks like Europe is creating a convenient system for its manufacturers, for its industries, so that they don't lose out in the competition with Asia and Russia, where labour and resource costs are lower.
>
> (R04, business media)

At the *routines level*, we identified certain rules that influence journalistic work and journalists' relationships with their sources. In their professional routines, journalists often have to adapt to the "rules of the game" or unspoken editorial policies. This "rules of the game" logic is most pronounced among those working in state-owned media, but it is also visible in other media when the owner or a major investor requests that criticism on certain issues should be avoided. One of the respondents considered adherence to these policies as a reflection of their professionalism:

> I'm a professional; I make my living from this, so I've never faced such a dilemma [conflict between my own position and that of an expert]. I'm a hired employee, and I have an employer who pays me. The employer has requirements: we do not criticize the decisions of the authorities and the laws that have been passed. As long as it's a debated idea, we can discuss it – say yes or no, be for or against it. But once it's accepted and implemented, then please stick to the rules.
>
> (R10, state-owned media)

Thus, the requirements of the owner/publisher not only influence the selection of sources, but also may shape the very notion of journalistic objectivity. By viewing playing by the rules as professionalism, a journalist gains an understanding of the limits of objectivity.

Time constraints are another important factor shaping journalistic practices, when the temporality of work in the newsroom limits the amount of time available to engage with sources. In practice, this means that journalists tend to contact those experts who are easier to reach and who respond more quickly. As such, the press services of federal state bodies are usually faster than those of regional authorities; larger ENGOs are more accessible than smaller ones; and well-known and public scientists are prioritised over less public ones. For federal state bodies, journalists have their own informal classification of "faster" and "slower" press services: "Rosleskhoz [the Federal Forestry Agency] was slow to comment. Rosprirodnadzor [the Federal Service for Supervision of Natural Resources] is not always on time. Maybe there just isn't enough staff, enough hands" (R08, commercial media).

Several respondents also mentioned that their newsroom closely follows metrics and news aggregators to decide what kind of stories to publish. This is a routine-level factor that stems from the structures of journalistic work, and journalists have to adapt to it by writing stories that will be widely read and shared, which also affects the sources they use.

At the *organisational level*, the ownership of the media organisation is the most salient factor influencing the selection of sources. Respondents had a clear idea about who they should represent first. For instance, an editor from a state-owned news agency pointed to the federal level and the biggest players:

> … [Our media] are still more tailored to the federal agenda and the agenda of government bodies. This means that [name of the media] first reports on what, for example, officials, or large industrialists from the RSPP [Russian Union of Industrialists and Entrepreneurs] have to say, and what people from organisations like Greenpeace, WWF and others say.[1]
>
> (R09, state-owned media)

A respondent working for an opposition investigative media outlet mentioned that open data serves as the initial source for their investigations, which are later substantiated with expert opinions. Independent media outlets potentially have access to a broader range of experts, from state officials to activists.

Many of our respondents, regardless of the type of media they were working for, cited Greenpeace as the best source of information, followed by the World Wildlife Fund (WWF). Both were very active in Russia before the full-scale invasion of Ukraine in February 2022, but have run into problems due to the increasing pressure on international organisations since the start of the invasion. The Office of the Russian Federation Prosecutor General declared Greenpeace (2023) an "undesirable" organisation in May 2023, followed by WWF (2023) in June 2023, resulting in the closure of both ENGOs' operations in Russia.

Respondents mentioned that Greenpeace was very effective in activating people, and without its significant resources, environmental activists were struggling to get their voices heard. This is particularly detrimental for non-state media, which find it difficult to obtain reliable expert information on environmental issues.

At the *social institutions level* we go beyond the media as organisations and analyse the relevant social institutions establishing and regulating the flow of information to the media.

Scientists in Russia are considered by journalists to be reliable and common sources. One interviewee stated, "In the sphere of ecology, the best sources are of course researchers, friends of science. They have devoted their lives to it and have probably done more than we ever can" (R13, commercial pro-state newspaper and online media). Many referred to scientists as approachable people who have interesting things to say and are eager to help. As one respondent noted, "I communicate mostly with researchers because they are usually well-educated, know everything very well, and have something to say" (R07, commercial news agency). Another respondent agreed that it was easy to talk to scientists, but noted that they are not used as the main source of information because "they rarely have any newsworthy stuff, just some general data" (R04, online business media). Not all researchers are easy to approach, but it is not difficult to find one who is: "There are always those who speak. Sure, there are also those who just don't like to speak, who don't see it as important. Then I just look for someone else who likes to talk" (R06, state-owned news agency).

Respondents also generally trusted the academic staff and their independence from state organs, albeit not always completely in the case of researchers at state-owned institutions. As one respondent remarked, "I haven't noticed any dependence... Well, I guess there is probably some dependency because the staff at state universities should follow the policies of their institutions, but it is not noticeable in conversations" (R07, commercial news agency).

However, two respondents voiced concerns about the independence and reliability of academic experts, citing a specific case where they noticed that scientists were not providing honest information. This concerned an incident in the Pacific Ocean near the Kamchatka Peninsula in September 2020 when a mass death of sea creatures was discovered, with up to 95% of marine life in the affected area dying. The vice president of Russia's Academy of Sciences told the media that this was due to natural causes, specifically the blooming of red algae (Phys.org, 2020). One of the respondents found this explanation implausible, however, and believed the cause was pollution from nearby industry.

Another respondent pointed out that Russian climatologists "do not criticize [President] Vladimir Putin or other official figures, and their statements clearly fall in line with those of the official authorities" (R18, environmental niche media). Yet another mentioned a case in which scientists from a state institution completely refused to speak to opposition media (R02, opposition online media). These examples demonstrate the influence at the level of social institutions, where generally reliable news sources – academic experts – may intentionally withhold information or frame messages falsely, thereby undermining the overall credibility of the academic society. According to our interviews, this may stem from the fact that state-funded research institutions are unable to act independently and therefore must, or choose to, align their positions with the interests of the state.

Corporations are less commonly used sources. While even opposition media can usually engage in dialogue with companies, corporations are the only sources that might have open conflicts with the media. For instance, our respondents described difficulties in establishing a dialogue with companies involved in the waste incineration business in the provinces. Local communities are typically against these projects, and companies react negatively to critical media coverage:

> We feel comfortable working with companies; they all talk to us. But sometimes there is a problem with waste incineration. One company that is going to build dozens of incineration plants in Russia has completely cut off contact with us. They think we're an opposition media outlet that's going against them. At first, they gave us comments, but after the first article was published, they said: "You're attacking us." I called them and tried to calm them down, but they said

> they wouldn't give us any more comments. All right, so we published the data on waste incineration without their commentary.
>
> (R08, niche online media)

At the level of social institutions, our research also revealed attitudes towards Greenpeace that differ considerably from those typically associated with the organisation internationally. On its homepage, Greenpeace International describes itself as an organisation that "uses peaceful, creative confrontation to expose global environmental problems,"[2] and that it invites "people out of their comfort zones to take courageous action."[3] Greenpeace is an atypical ENGO in that it has been organised globally from its inception and operates with a strongly hierarchical top-down structure, where "professional activists do the campaigning and ordinary members raise funds" (Eden, 2006). Its actions have sparked controversies, and Greenpeace has an established reputation as an activist organisation.

However, many of our respondents referred to Greenpeace as an expert organisation, where many people do not think of themselves as environmental activists. This expert positioning is probably one of the main reasons why many Russian environmental journalists relied on Greenpeace as their main source of information before it was closed in Russia in summer 2023.

Three out of four respondents interviewed before the full-scale invasion of Ukraine (R01, opposition online media; R02, opposition online media; R03, business media) praised Greenpeace for its openness, accessibility, and extensive knowledge. Unlike with state officials, with Greenpeace journalists did not have to work with press services; instead, they were directly connected with experts in the relevant field. It is striking how many Russian environmental journalists seemed to treat Greenpeace's experts not as activists but as the best and most reliable environmental experts available:

> I seek comments from Greenpeace Russia quite often, mainly because there are people there who used to work for the Ministry of Natural Resources and similar organisations. This is why they understand how things work, follow all the agendas and are very knowledgeable about the legislation. ... From what I've gathered, not everyone at Greenpeace positions themselves as an activist. ...some of them clearly position themselves as activists and engage in all sorts of actions, but I don't talk to them that often. For me, it's more important that a person is an expert, and there are quite a few such experts out there [in Greenpeace].
>
> (R03, business media)

Greenpeace's positioning can be seen as a purposeful adaptation strategy by NGOs in response to state pressure, shifting from civic and political activism to providing consultancy and expert services (Tysiachniouk et al., 2023). At the same time, this is in stark contrast to how organisations and people considered activists are often treated, namely as not the best sources of information:

> I communicate with activists to a lesser extent because their actions are usually seasonal, taking place only in the summer. They are just not around that much. And, as a rule, there's no need for personal communication with them; everything can be handled effectively through their press releases.
>
> (R07, commercial news agency)

Finally, governmental sources are engaged by all media due to their hub position in providing access to data and information on environmental issues. However, access to this data is unequal and tends to privilege state-owned and/or pro-state media:

> I receive comments [from the Ministry of Natural Resources and Environment] directly. Well, not exactly directly – there are always press secretaries involved. In principle, press secretaries are convenient; I like them. I wouldn't say it's difficult to obtain a comment. It's more about what kind of comment you get. Well, at least I'll get something. In principle, this is the advantage of working at a state-owned agency: we always get a comment. But colleagues who work at less important publishing houses or non-state ones might get nothing at all.
>
> (R06, state-owned news agency)

As Anderson (2017) notes, PR practitioners play a transformative role in journalist-source relations, blurring the traditional journalistic practice of accessing experts. According to our respondents, PR departments of state bodies are becoming the contact point for journalists. While they speed up the process of obtaining responses, they also decrease the likelihood of acquiring exclusive information first hand (Samoilenko & Erzikova, 2021). In the case of autocracies, they also serve as additional filters, determining which media outlets receive information and what kind of information is published.

The *social system level* engages with broader societal forces and values (ideological, cultural, political, and economic) and provides a deeper understanding of the relationship between journalists and their sources.

The main message expressed by our respondents was the issue of information being withheld or hidden, a feature often associated with autocracies (Guriev & Treisman, 2019). For instance, some ministries may not respond at all, or if they do, it can take weeks. In addition, as discussed above, it might not be possible to get in direct contact with state experts themselves, with responses typically coming from press secretaries. Those who commented on the quality of the data provided by state bodies generally described it as having very little value. One respondent considered that the Ministry of Natural Resources and Environment, for example, is not an independent actor, and its staff are unable to express their true opinions. Others blamed the problems on deep-seated bureaucracy, which makes working with state institutions challenging. On the other hand, one respondent acknowledged that two organisations at least maintain communication: the Federal Service for Supervision of Natural Resources (*Rosprirodnadzor*), and the Federal Service for Environmental, Technological and Nuclear Supervision (*Rostekhnadzor*).

Respondents also pointed out that a substantial part of the data related to environmental and climate issues is not being collected at all.

The features of journalist-state source relations in Russia are aligned with the current political regime and the role of the media within it. Guriev and Treisman (2019) describe contemporary Russia as an informational autocracy where the covert subversion of political institutions is concealed behind a democratic façade. The key to such regimes is the manipulation of information to ensure state dominance in political communications with ordinary citizens. Following this logic, government bodies are expected to demonstrate their commitment to formal democratic institutions, such as public accountability, by appearing open to the media. At the same time, monopolisation of the media agenda is a crucial element for boosting the regime's popularity and eliminating threats (Guriev & Treisman, 2019). As such, state bodies, as media sources, tend to establish reliable and open relations with journalists while carefully filtering the data they provide to journalists at the same time. The development of press services in recent years (Samoilenko & Erzikova, 2021), coupled with simultaneous difficulties in gaining direct access to state officials, has created a situation whereby journalists cannot complain about a lack of access to government data, but the data they receive is filtered according to the state's needs.

Conclusions

This chapter explores how journalist-source relations are structured in environmental reporting within the Russian media. Utilising the HOI model, our analysis unpacks the way in which sourcing is influenced not only by journalistic culture, but also by broader forces such as social institutions and the social system at large. This perspective allows for a more nuanced understanding of how the environmental agenda is constructed in an authoritarian state when media freedoms are restricted.

Our findings uncover a multi-layered picture in which personal professional practices, together with organisational or institutional norms and societal values at large, shape the process of sourcing in journalistic work. As we discovered, when selecting sources, journalists rely on an understanding of professionalism as a balancing act between adhering to the "rules of the game" of a newsroom and the media environment at large, while also exercising their own agency and gatekeeping. As such, journalists tend to routinely contact government sources, but also seek to verify or corroborate the data with input from scientific or activist experts. However, while government sources are a must in the authoritarian context of the Russian media environment, the other sources and how they are used depend on the journalist's experience, personal opinions, temporality of the work, and the media outlet they work for. The recent shutdown of leading ENGOs (notably Greenpeace and WWF) and the mass exodus of independent media (Dovbysh & Rodina, 2022) raise questions about what alternatives to state-controlled narratives will emerge on the environmental and climate agenda.

Despite the dominance of government sources, journalists perceive scientists as the most reliable, substantive, and accessible sources, especially for opposition media. At the same time, the Covid-19 pandemic and the subsequent war in Ukraine created crisis conditions that transformed science-focused journalism from a soft news beat into one that dealt with highly politicised events (Borissova Saleh, 2024). When it comes to environmental and climate reporting, which occupies an intermediary position in Russian newsrooms between the science, social, and business departments, we identified an alternative strategy. The journalists tend to depoliticise the most "politically loaded" beats of climate crisis or other environmental problems by framing them as purely scientific puzzles (Dovbysh & Perkiömäki, 2024).

Although the politicisation of the environmental agenda and the dominance of governmental sources are global trends, we identified unique characteristics of these processes within the Russian media environment. The limited diversity of voices among political elites, combined with the state pressure on alternative voices such as ENGOs and scientists, leads to the unification and formalisation of the environmental agenda. Efforts by PR departments of state bodies to produce acceptable data for the media, along with simultaneous control over access to spokespersons, further contribute to this tendency. Discussing the journalistic sourcing of climate change reporting in China, Guo et al. (2023, p. 514) point out that this leads to the framing of climate change as "a political goal to achieve for the good of the government's image." We observe similar trends in Russia, where environmental and climate reporting either indicate the authorities' activities or represent a neutral and detached discussion of ongoing crises. As the analysis above demonstrates, the personal engagement of journalists and their dedication to the topic could counteract these trends. However, the increasing pressure on independent journalism in Russia poses a significant threat to more involved and dedicated environmental journalism.

Acknowledgements

This research has been conducted as part of the project: "FLOWISION: Best from both worlds – enhancing energy transition in Russia and Finland by making resource flows visible," funded by the Kone Foundation. See https://flowision.fi

Notes

1 With this quotation the respondent stresses out the significance of large, powerful sources rather than the fact that governmental sources, industrial sources and ENGOs are of equally importance for them. At the time of interview, Greenpeace and WWF operated in Russia without restrictions and obtained the positions of powerful actors in the area, and that is why the respondent mentions them.

2 See www.greenpeace.org/international/about/values/

3 See www.greenpeace.org/international/about/

References

Anderson, A. (2017). Source influence on journalistic decisions and news coverage of climate change. In H. von Storch (Ed.), *Oxford research encyclopedia of climate science* (pp. 1–34). Oxford University Press.

Ashe, T. & Poberezhskaya, M. (2022). Russian climate scepticism: an understudied case. *Climatic Change, 172*(3–4): article number 41. https://doi.org/10.1007/s10584-022-03390-3

Atton, C. & Wickenden, E. (2005). Sourcing routines and representation in alternative journalism: a case study approach. *Journalism Studies*, *6*(3), 347–359.

Beuerle, B. (2023). From continuity to change: Soviet and Russian government attitudes on climate change (1989–2009). *Climatic Change 176*, article number 36. https://doi.org/10.1007/s10584-023-03488-2

Borissova S.A. (2024) "The death of experts": sourcing science journalism in a state-controlled media environment. In F. Mellor (Ed.), *Insights on science journalism* (pp. 121–138). Routledge.

Bouzarovski, S. & Bassin, M. (2011). Energy and identity: imagining Russia as a hydrocarbon superpower. *Annals of the Association of American Geographers*, *101*(4), 783–794. http://dx.doi.org/10.1080/00045608.2011.567942

Brossard, D., Shanahan, J., & McComas, K. (2004). Are issue-cycles culturally constructed? A comparison of French and American coverage of global climate change. *Mass Communication & Society*, *7*(3), 359–377. https://doi.org/10.1207/s15327825mcs0703_6

Carlson, M. (2009). Dueling, dancing, or dominating? Journalists and their sources. *Sociology Compass*, *3*(4), 526–542. https://doi.org/10.1111/j.1751-9020.2009.00219.x

Comfort, S.E., Tandoc, E., & Gruszczynski, M. (2020). Who is heard in climate change journalism? Sourcing patterns in climate change news in China, India, Singapore, and Thailand. *Climatic Change*, *158*, 327–343. https://doi.org/10.1007/s10584-019-02597-1

Davydova, A. (2020). Environmental journalism in Russia. In D.B. Sachsman & J.A. Myer Valenti (Eds.), *Routledge handbook of environmental journalism* (pp. 246–249). Routledge.

Dovbysh, O., & Perkiömäki, M. (2024). (De)politicization of the environmental agenda in Russian media. *Journalism*. https://doi.org/10.1177/14648849241284338

Dovbysh, O., & Rodina, E. (2022). Russian independent journalism in exile: In search of relevance and resistance, *Russia.Post*. Retrieved June 12, 2024, from https://russiapost.info/society/journalism_in_exile

Duan, R., & Takahashi, B. (2017). The two-way flow of news: a comparative study of American and Chinese newspaper coverage of Beijing's air pollution. *International Communication Gazette*, *79*(1), 83–107. https://doi.org/10.1177/1748048516656303

Eden, S. (2006). Greenpeace. *New Political Economy*, *9*(4), 595–610. https://doi.org/10.1080/1356346042000311191

Ejaz, W., Ittefaq, M., & Arif, M. (2022). Understanding influences, misinformation, and fact-checking concerning climate-change journalism in Pakistan. *Journalism Practice*, *16*(2–3), 404–424. https://doi.org/10.1080/17512786.2021.1972029

Esli byt' tochnym [To be precise]. (2023). За последние два года 44 органа власти удалили со своих сайтов почти 500 датасетов. [Over the past two years, 44 authorities have removed almost 500 datasets from their websites]. Retrieved March 12, 2024, from https://tochno.st/materials/za-poslednie-dva-goda-44-organa-vlasti-udalili-so-svoix-saitov-pocti-500-datasetov-itogovoe-obnovlenie-trekera-otkrytyx-dannyx-ot-esli-byt-tocnym

FOM. (2024, 20 June). *Climate change [Изменения климата]*. Public Opinion Foundation. Retrieved June 26, 2024, from https://fom.ru/Obraz-zhizni/15033

Greenpeace. (2023, 19 May). Environmental protection is 'undesirable' in Russia. Retrieved March 12, 2024, from www.greenpeace.org/international/press-release/59745/environmental-protection-undesirable-russia-greenpeace/

Guo, J., Huang, X., & Fang, K. (2023). Authoritarian environmentalism as reflected in the journalistic sourcing of climate change reporting in China. *Environmental Communication*, *17*(5), 502–517. https://doi.org/10.1080/17524032.2023.2223774

Guriev, S., & Treisman, D. (2019). Informational autocrats. *Journal of Economic Perspectives*, *33*(4), 100–127. https://doi.org/10.1257/jep.33.4.100

Gustafson, T. (2021). *Klimat. Russia in the age of climate change*. Harvard University Press. https://doi.org/10.2307/j.ctv1xtwqdr

Huan, C. (2024). Politicized or popularized? News values and news voices in China's and Australia's media discourse of climate change. *Critical Discourse Studies*, *21*(2), 200–217. https://doi.org/10.1080/17405904.2023.2200194

Kvale, S. (1996). *InterViews: an introduction to qualitative research interviewing*. Sage.

Laursen, B., & Trapp, N.L. (2021). Experts or advocates: shifting roles of central sources used by journalists in news stories? *Journalism Practice*, *15*(1), 1–18. https://doi.org/10.1080/17512786.2019.1695537

Phys.org. (2020, 12 October). Kamchatka marine life death caused by algae: Russian scientist. Retrieved March 11, 2024, from https://phys.org/news/2020-10-kamchatka-marine-life-death-algae.html

Poberezhskaya, M. (2015). Media coverage of climate change in Russia: governmental bias and climate silence. *Public Understanding of Science, 24*(1), 96–111. https://doi.org/10.1177/0963662513517848

Poberezhskaya, M. (2018). Traditional media and climate change in Russia A case study of *Izvestiia*. In M. Poberezhskaya & T. Ashe (Eds.), *Climate change discourse in Russia: Past and present* (pp. 64–79). Routledge. https://doi.org/10.4324/9781351028660

Reese, S.D., & Shoemaker, P.J. (2016). A media sociology for the networked public sphere: the hierarchy of influences model. *Mass Communication and Society*, *19*(4), 389–410. https://doi.org/10.1080/15205436.2016.1174268

Reich, Z., & Barnoy, A. (2021). Disagreements as a form of knowledge: how journalists address day-to-day conflicts between sources. *Journalism*, *22*(4), 882–900. https://doi.org/10.1177/1464884918825007

Sakwa, R. (2021). Greater Russia: is Moscow out to subvert the West? *International Politics*, *58*(3), 334–362. https://doi.org/10.1057/s41311-020-00258-0

Samoilenko, S.A., & Erzikova, E. (2021). Public relations in Russia. In M.C. Minielli, M.N. Lukacovic, S.A. Samoilenko & M.R. Finch (Eds.), *Media and public relations research in post-socialist societies*. Rowman Littlefield.

Schäfer, M.S., & Painter, J. (2021). Climate journalism in a changing media ecosystem: assessing the production of climate change-related news around the world. *Wiley Interdisciplinary Reviews: Climate Change*, *12*(1). https://doi.org/10.1002/wcc.675

Shoemaker, P., & Reese, S.D. (2014). *Mediating the message in the 21st century: a media sociology perspective*. Routledge.

Takahashi, B., Huang, K., Fico, F., & Poulson, D. (2017). Climate change reporting in Great Lakes region newspapers: a comparative study of the use of expert sources. *Environmental Communication*, *11*(1), 106–121. http://dx.doi.org/10.1080/17524032.2016.1220967

Tuchman, G. (1978). *Making news: a study in the construction of reality*. Free Press.

Tynkkynen, V.-P. (2019). *The energy of Russia*. Edward Elgar Publishing.

Tynkkynen, V.-P. & Tynkkynen, N. (2018). Climate denial revisited: (Re)contextualising Russian public discourse on climate change during Putin 2.0. *Europe-Asia Studies, 70*(7), 1103–1120. https://doi.org/10.1080/09668136.2018.1472218

Tysiachniouk, M.S., Tulaeva, S.A., Kotilainen, J., & Henry, L.A. (2023). Liberal spaces in an illiberal regime: environmental NGOs, state sovereignty and the struggle for nature. *Territory, Politics, Governance*, 1–20. http://dx.doi.org/10.1080/21622671.2023.2186942

Vu, H.T., Liu, Y., & Tran, D.V. (2019). Nationalizing a global phenomenon: a study of how the press in 45 countries and territories portrays climate change. *Global Environmental Change, 58*. https://doi.org/10.1016/j.gloenvcha.2019.101942

Weiner, D.R. (1999). *A little corner of freedom: Russian nature protection from Stalin to Gorbachëv*. University of California Press.

Wilson Rowe, E. (2009). Who is to blame? Agency, causality, responsibility and the role of experts in Russian framings of global climate change. *Europe-Asia Studies*, *61*(4), 593–619. https://dx.doi.org/10.1080/09668130902826154

WWF. (2023, 22 June). WWF International statement on "undesirable" characterisation in Russia. Retrieved August 1, 2024, from https://updates.panda.org/wwf-international-statement-on-undesirable-characterisation-in-russia

Yagodin, D. (2017). Climate change communication in Russia. *Oxford research encyclopedia of climate science*. Oxford University Press.

13 Conclusion

From constructing facts to constructing expertise and trust?

Anna Rantasila, Anna Maria Jönsson, and Mette Marie Roslyng

In the Introduction to this volume, we set out to answer the following question: *Whose truth? Questions of definition in the age of contested science*. As the 12 chapters in three thematic sections attest, the answer varies considerably depending on the context in which the question is asked. One of the recurring themes across the chapters is the clashing of different forms of institutional science and technology expertise (Jönsson et al., 2025). News media and social media contribute in various ways to the construction of knowledge and counter-knowledge in all the formats and practices discussed across the chapters. News media may prioritise and highlight certain forms of institutionalised knowledge while down-playing alternative citizen voices as shown by Mäkinen (2025). Indeed, several political actors that lean more into counter-knowledge discourse position themselves as marginalised and less powerful in defining the truths and facts in mainstream media, as e.g., Vécsey (2025), Agustín and Jerne (2025), and Demata (2025) point out in their analyses. Furthermore, argumentation styles and language of the scientific establishment can be appropriated by representatives of counter-knowledge and mainstream news media alike to support their own claims, as Abalo (2025) and Demata (2025) illustrate in their chapters. Finally, adding the construction of literary scientific imaginaries, as in cli-fi (often dystopian) narratives, can broaden the scope of science representation (Andersen, 2025). These phenomena can be understood as a part of wider development where the boundaries of scientific and technological expertise are renegotiated and challenged by various actors that come from outside of the techno-scientific establishment (Uutela & Väliverronen, 2024).

According to Uutela and Väliverronen (2024), both mainstream and social media in the hybrid media system (Chadwick, 2013) play a significant role in both legitimising and challenging expertise. This dynamic is reflected in chapters by Jönsson (2025), Mäkinen (2025), Larsen and Roslyng (2025), and Sarlos (2025). These chapters emphasise both how media frame science and technology-related topics and different types of expertise, and how this may influence public perceptions of these topics. These chapters also make visible a crucial tension in Science and Technology Studies, described by Krick and Meriluoto (2022) as tension between epistemic and democratic approaches to knowledge: Should one emphasise the special quality of scientific expertise and acknowledge the limits of

DOI: 10.4324/9781003479550-16

lay knowledge, or should one embrace the lay audiences potential and encourage the experts to engage with a wider range of publics? We maintain that these stances are not mutually exclusive but must be seen as two ends of a spectrum that is often very contextual. For example, a nuclear physicist holds profound knowledge on radiation, but the concerns of lay people living next to a nuclear power plant should not be dismissed as ignorance. To paraphrase Durant (2023), scientific experts need reflexivity that enables them to provide information based on their knowledge, while simultaneously engaging with the ethical, social and political aspects of their work. Media representations of science have an important role to play in this process and can contribute both to the hindrance and to the enabling of scientific reflexivity.

Contested facts and construction of populist arguments

The call for reflexivity, as expressed by Durant (2023) and Wynne (2014), becomes highly visible in debates around issues that involve science and technology are also tied to political populism. As Agustín and Jerne (2025) point out in their chapter on green populism, various stakeholders engage in populist tactics to gain legitimacy for their claims in the public discourse. Or as Demata (2025) illustrates in his chapter, conspiratorial media figures may use visual and rhetorical tactics of validation of information that highly resemble those of established science and mainstream media, producing what Demata calls "epistemic masquerade". In other words, these figures use techniques resembling scientific and journalistic argumentation to discredit the institutions they borrow their tools from.

Moreover, how different fields of science are discussed in the media can also be tied to populist arguments around public funding and sociocultural value of research, as Jönsson's (2025) chapter illustrates through an analysis of Swedish mainstream media discourse about science and research. Debates about social and cultural significance of STEM fields over humanities and social sciences, for example, have been connected to the rise of right-wing populism and so-called "culture wars" between socially conservative and liberal stakeholders, particularly in the US and the UK (e.g., Brint et al., 2024; Clemens & Hochmuth, 2024). These contestations can be further connected to the wider phenomenon of political polarisation and movements that either deliberately or as an unintended consequence weaken trust in democratic processes and institutions (e.g., Schatto-Eckrodt et al., 2024; Alinejad & Van Dijk, 2023; Carlson, 2018; Fawzi, 2018; Holt, 2018). As Dovbysh and Perkiömäki (2025) point out in their chapter, this development is particularly concerning in societies such as Russia and other authoritarian regimes, where media, scientists, and civil society are all under pressure from an undemocratic regime. It is in these debates and contestations where scientific experts' ability to reflect upon the ethical and political implications of their work becomes crucial.

In addition to being tied to political power and legitimacy in society, contestations over truth and facts are also tied to emotions, which also can be used to fuel populist sentiment (Wahl-Jorgensen, 2019). As Rantasila (2025) points out in her chapter, trust in science and technology requires careful management, and broken trust can give rise to a wide range of emotions, including fear and anger. Moreover, trust in science and scientific expertise can be described as epistemic trust that can be nurtured through education about science, as well as through transparency and accountability of the scientific community (Jönsson et al., 2025). Here, social media plays a significant role, as it provides an opportunity for science communicators to reach and engage with their audiences in multiple ways that differ from the traditional understanding of scientific experts as detached (Durant, 2023).

However, as several chapters in this volume highlight (Sarlos, 2025; Vécey, 2025; Larsen & Roslyng, 2025), the interplay of social media and mainstream media in the hybrid media system means that the representatives of the scientific establishment must compete for attention of the multiple audiences with lay experts and holders of counter-knowledge. This competition for attention can fuel polarisation and antagonistic relationships between members of the scientific community and holders of counter-knowledge (Collins et al., 2020). This leads us toward some final remarks and possible paths forward, regarding the role of scientific expertise and construction and contestation of truth.

Concluding remarks and paths forward

With the chapters of this volume, we have explored the contestations of discursive power that entail science and environmental communication. The chapters and their empirical cases paint a vivid picture of the field, emphasising how issues related to science and technology are deeply embedded within the sociopolitical and cultural framework of contemporary societies. We argue that these contestations of truth, when understood as differences in opinion and exercises in civic debate, are a part of contemporary democracies where a broad variety of thought and belief coexist. These claims are often also connected to genuine and legitimate concerns about trustworthiness of science and technology, as chapters by Demata (2025), Larsen and Roslyng (2025), and Mäkinen (2025), as well as previous research (e.g., Oikkonen, 2017) demonstrate. However, Collins and colleagues (2020) quite provocatively state that scientists, and social scientists in particular, should more boldly stand up to the holders of counter-knowledge and become more empowered as holders of expert knowledge. In order to do this, Collins and colleagues (2020, 56) suggest that the scientific community should focus more on articulating their position as holders of expert knowledge and embrace their role as a vital part of a pluralist contemporary democracy. They argue that adopting this stance does not exclude taking seriously the concerns that often foster seeds of mistrust of science and counter-knowledge, as it calls for reflexivity (Collins et al., 2020; Wynne 2014). We believe that through the analysis of counter-knowledge arguments and scientific expertise alike, this volume demonstrates just that.

Examining all chapters of this volume together, we suggest that in addition to dissemination and definition of facts, communication about science, technology, and environmental issues is also about the feelings and emotions these issues invoke. As members of the scientific community, we suggest reflection on how complex sociotechnical issues, such as global pandemics, climate change, and environmental pollution are communicated in a way that does not dismiss people's concerns. This way, we suggest that the scientific community and communicators of science can answer to the call made by Collins and his colleagues (2020; 2023), and to build trust in science in a way that is contributive to both the role of experts and to pluralist democracy (Durant, 2023).

References

Abalo, E. (2025) The scientification of risks and the risks of scientification: insights from the coverage of artificial turf pitches as microplastic pollutants in Sweden. In M.M. Roslyng, A. Rantasila, & A.M. Jönsson, (Eds.), *Communicating science, climate change and the environment in hybrid media: constructed facts, contested truths* (pp. 15–34). Routledge.

Agustín, O.G. & Jerne, I. (2025) Green populism. Counterpublics and the formation of counterknowledge. In M.M. Roslyng, A. Rantasila & A.M. Jönsson, (Eds.), *Communicating science, climate change and the environment in hybrid media: constructed facts, contested truths* (pp. 70–88). Routledge.

Alinejad, D. & Van Dijk, J. (2023) Climate communication: how researchers navigate between scientific truth and media publics. *Communication and the Public*, *8*(1), 29–44. https://doi.org/10.1177/20570473221138612

Andersen, G. (2025). Cli-fi and five narratives of future warming. In M.M. Roslyng, A. Rantasila & A.M. Jönsson (Eds.), *Communicating science, climate change and the environment in hybrid media: constructed facts, contested truths* (pp. 57–69). Routledge.

Brint, S., Webb, M. & Fields, B. (2024). An Uneasy Peace: how STEM progressive, traditionalist, and bridging faculty understand campus conflicts over diversity, anti-racism, and free expression. *Minerva, 62*, 339–372. https://doi.org/10.1007/s11024-023-09517-9

Carlson, M. (2018) The information politics of journalism in a post-truth age. *Journalism Studies*, *19*(13), 1879–1888. https://doi.org/10.1080/1461670X.2018.1494513

Chadwick, A. (2013) *The hybrid media system: Politics and power*. 1st edn, Oxford Studies in Digital Politics. Oxford University Press. https://doi.org/10.1093/acprof:oso/9780199759477.001.0001

Clemens, M. & Hochmuth, C. (2024). Political speech on campus: shifting the emphasis from "if" to "how". *Minerva, 62*, 373–396. https://doi.org/10.1007/s11024-024-09525-3

Collins, H., Evans, R., Durant, D. & Weinel, M. (2020) *Experts and the will of the people. society, populism and science*. Palgrave Pivot.

Collins, H., Evans, R., Durant, D. & Weinel, M. (2023). The third wave and populism: scientific expertise as a check and balance. In G. Eyal & T. Medvetz (Eds.), *The Oxford handbook of expertise and democratic politics* (pp. 52–75). Oxford University Press. https://doi.org/10.1093/oxfordhb/9780190848927.013.3

Demata, M. (2025). Fighting (for) truth? Alex Jones, the WHO and the legitimization of conspiracy discourse. In M.M. Roslyng, A. Rantasila & A.M. Jönsson (Eds.), *Communicating science, climate change and the environment in hybrid media: constructed facts, contested truths* (pp. 91–111). Routledge.

Dovbysh, O. & Perkiömäki, M. (2025). Journalists/sources relations in Russian environmental journalism. In M.M. Roslyng, A. Rantasila & A.M. Jönsson (Eds.), *Communicating science, climate change and the environment in hybrid media: constructed facts, contested truths* (pp. 216–232). Routledge.

Durant, D. (2023). Are honest brokers good for democracy? *Social Epistemology*, *37*(3), 276–289. https://doi.org /10.1080/02691728.2022.2139166

Fawzi, N. (2018). Untrustworthy news and the media as "enemy of the people?" How a populist worldview shapes recipients' attitudes toward the media. *The International Journal of Press/Politics*, *24*(2), 146–164. https://doi.org/10.1177/1940161218811981

Holt, K. (2018) Alternative media and the notion of anti-systemness: towards an analytical framework. *Media and Communication*, *6*(4), 49–57. https://doi.org/10.17645/mac.v6i4.1467

Jönsson, A.M. (2025) Mediated science and issues of public knowledge and trust. In M.M. Roslyng, A. Rantasila & A.M. Jönsson, (Eds.), *Communicating science, climate change and the environment in hybrid media: constructed facts, contested truths* (pp. 153–175). Routledge.

Jönsson, A.M., Roslyng, M.M. & Rantasila, A. (2025). Introduction: contesting truths in science and environment communication. In M.M. Roslyng, A. Rantasila & A.M. Jönsson (Eds.), *Communicating science, climate change and the environment in hybrid media: constructed facts, contested truths* (pp. 1–14). Routledge.

Krick, E. & Meriluoto, T. (2022) The advent of the citizen expert: democratising or pushing the boundaries of expertise?, *Current Sociology*, *70*(7), 967–973. https://doi.org/10.1177/00113921221078043

Larsen, G. & Roslyng, M.M. (2025) Knowledge and counter-knowledge: the discursive-linguistic construction of facts in vaccination debates. In M.M. Roslyng, A. Rantasila & A.M. Jönsson (Eds.), *Communicating science, climate change and the environment in hybrid media: constructed facts, contested truths* (pp. 112–130). Routledge.

Mäkinen, M. (2025). Citizen activists or pandemic deniers? Alternative voices in the Finnish journalistic media during the COVID-19 pandemic. In M.M. Roslyng, A. Rantasila & A.M. Jönsson (Eds.), *Communicating science, climate change and the environment in hybrid media: constructed facts, contested truths* (pp. 131–150). Routledge.

Oikkonen, V. (2017). Affect, technoscience and textual analysis: interrogating the affective dynamics of the Zika epidemic through media texts. *Social Studies of Science*, *47*(5), 681–702. https://doi.org./10.1177/0306312717723760

Rantasila, A. (2025) Constructing trust through affective discipline: Finnish nuclear energy experts and the Fukushima Daiichi disaster. In M.M. Roslyng, A. Rantasila & A.M. Jönsson (Eds.), *Communicating science, climate change and the environment in hybrid media: constructed facts, contested truths* (pp. 176–193). Routledge.

Sarlos, G. (2025) Nuclear stories in the news media. Filtering and altering of expert views. In M.M. Roslyng, A. Rantasila & A.M. Jönsson (Eds.), *Communicating science, climate change and the environment in hybrid media: constructed facts, contested truths* (pp. 194-215). Routledge.

Schatto-Eckrodt, T., Clever, L., & Frischlich, L. (2024). The seed of doubt: examining the role of alternative social and news media for the birth of a conspiracy theory. *Social Science Computer Review*, *42*(5), 1160–1180. https://doi.org./10.1177/08944393241246281

Uutela, E. & Väliverronen, E. (2024). Redefining expertise: how proponents and opponents of alternative therapies evaluate expertise. *Science as Culture*. DOI: 10.1080/09505431.2024.2416675

Vécsey, V. (2025). Web of denial. Climate change denial discourse on Instagram. In M.M. Roslyng, A. Rantasila & A.M. Jönsson (Eds.), *Communicating science, climate change and the environment in hybrid media: constructed facts, contested truths* (pp. 35–56). Routledge.

Wahl-Jorgensen, K. (2019). *Emotions, media, and politics. Contemporary political communication*. Polity.

Wynne, B. (2014). Further disorientation in the hall of mirrors. *Public Understanding of Science*, *23*(1), 60–70. doi:10.1177/0963662513505397.

Index

Note: Tables are indicated by bold.

Kolibri 360 GmbH
Ferdinand-Jühlke-Straße 7
99095 Erfurt, Deutschland
produktsicherheit@kolibri360.de